Faraday Reimagined

Sandeep Chavan

Published by Gyrus Vision, 2026.

FARADAY REIMAGINED

First edition. January 18, 2026.

ISBN: 979-8233090721

Written by Sandeep Chavan.

Table of Contents

To those who prefer clarity over certainty,and restraint over invention.

When explanation outruns necessity,language begins to invent.

Sandeep Chavan

Opening Note

In Gratitude to Curiosity

This book began, as many meaningful things do, not in a laboratory or a lecture hall, but in a conversation—slightly chaotic, mildly provocative, and unexpectedly productive. A casual challenge on a physics post, a surname borrowed with confidence, and a familiar question about *qualifications*. Instead of closing the discussion, it opened something far more valuable: *electromagnetism itself.*

So, before anything else, let me say this plainly—*thank you.* Thank you to curiosity, even when it arrives disguised as skepticism. Thank you to discomfort, because it forces clarity. And thank you to the quiet realization that real physics does not advance through credentials alone, but through *engagement with nature.*

That realization takes us directly to **Michael Faraday**.

Faraday was not armed with advanced mathematics. He did not belong to the elite academic structures of his time. What he possessed instead was something far more enduring: an uncompromising relationship with observation. He trusted what nature revealed when systems changed, when equilibrium was disturbed, when configurations refused to stay still. Faraday did not chase formulas; he followed *change.*

Modern physics celebrates his results, teaches his laws, and builds industries on his discoveries. Yet, somewhere along the way, we quietly replaced his intuition with abstractions so powerful that they began to feel real in themselves. Fields became entities. Flux lines became things. Language became mechanism.

This book is not an attempt to diminish that progress. Engineering achievements, mathematical formalism, and technological mastery deserve their place. But understanding is a different currency altogether. And understanding demands that we periodically return to first principles—not to reject what we know, but to ask whether we still know *why*.

The **Universal Energy Dynamics (UED)** framework emerged from precisely this kind of questioning. It does not seek to overthrow experimental physics. It does not rewrite outcomes. What it does instead is re-examine *ontology*—what we assume exists, what we treat as cause, and what we quietly accept as explanatory shortcuts.

Through the lens of UED, Faraday's work appears not as an incomplete prelude to Maxwell, but as a remarkably coherent intuition that was later translated into a language he himself never claimed to own. His "lines of force" were never meant to be physical threads. His induction experiments were never about invisible cutting. They were about *energy reorganizing itself when equilibrium is disturbed.*

Seen this way, Faraday does not belong only to the history of electromagnetism. He belongs to its future.

This book is written neither as a biography nor as a textbook. It is a reinterpretation—calm, respectful, and unapologetically independent. Each chapter revisits Faraday's key inventions and insights, not to modernize them, but to *restore their conceptual integrity* using a framework that emphasizes configuration, boundary, delta, and equilibrium.

You will not find heavy mathematics here. That absence is intentional. Faraday himself would have approved. Equations are powerful tools, but tools are not truths. When understanding matures, mathematics follows naturally. When mathematics leads without understanding, language quietly replaces logic.

If you are a student, this book invites you to see electromagnetism without fear. If you are an engineer, it invites you to look beneath what already works. If you are a thinker, it invites you to question what has become comfortable.

And if you arrived here because of a disagreement, a comment, or a challenge—welcome. You are exactly where Faraday would have wanted the conversation to go.

Electromagnetism did not begin with formulas. It began with curiosity about change.

This book simply continues that conversation.

Preface

Why Faraday, Why Now

Every generation inherits physics as a finished structure—polished laws, standardized explanations, and formulas that work so reliably that questioning them feels unnecessary, even irresponsible. Electromagnetism, perhaps more than any other domain, gives this impression of completeness. Motors spin, transformers hum, power grids operate with precision. From an engineering standpoint, the story appears closed.

And yet, understanding is not the same as utility.

This book begins with a simple but uncomfortable observation: *we use Faraday everywhere, but we no longer think like him.*

Faraday did not approach nature armed with equations or institutional authority. He approached it with patience, honesty, and a refusal to explain change using static ideas. He trusted experiments over language, patterns over prestige, and intuition over inheritance. His genius was not that he discovered electromagnetism as we now describe it, but that he *saw* something deeper at a time when the conceptual tools to describe it did not yet exist.

Faraday lived in an era when electricity and magnetism were fragmented curiosities. By the time we encounter them today, they are unified, mathematized, and industrialized. But in that very success lies a subtle loss. Somewhere between Faraday's laboratory and modern textbooks, intuition was replaced by abstraction, and abstraction gradually hardened into ontology. What began as visual aids and explanatory metaphors—fields, lines, flux—quietly became treated as physical actors in their own right.

This shift was not a mistake. It was a necessity of its time. Mathematics needed stable entities to operate on. Engineering needed predictive formalisms. Progress demanded compression. But compression always discards nuance.

Why revisit Faraday now?

Because we have reached a stage where the limits are no longer technological, but conceptual. We can build extraordinarily efficient systems, yet we struggle to explain them without leaning on language that even experts admit is metaphorical. We teach students to calculate before they are allowed to understand. We answer "how much" with confidence, but hesitate when asked "what is really happening".

At the same time, physics itself is facing deeper discomforts—fragmentation across scales, interpretational overload, and increasing dependence on mathematical patchwork to reconcile incompatible frameworks. These are not failures of intelligence. They are signals that our ontological assumptions deserve re-examination.

This is where the Universal Energy Dynamics (UED) framework enters—not as a rebellion, but as a restoration.

UED does not challenge Faraday's observations. It does not dispute experimental laws or engineering outcomes. Instead, it asks a quieter question: *What if Faraday's intuition was pointing toward energy reconfiguration and equilibrium-seeking behavior, rather than invisible entities acting at a distance?* What if induction, motion, and coupling were never about "fields doing work," but about structured energy responding to change?

When viewed through this lens, Faraday's work appears remarkably modern. His emphasis on change, boundary, geometry, and configuration aligns more closely with contemporary systems thinking than with the force-centric language often attributed to him retroactively. He was not primitive Maxwell; he was something different altogether—an observer of dynamic order.

Why now?

Because today we have something Faraday did not:

- a mature understanding of systems
- cross-disciplinary insights into equilibrium, stability, and dynamics
- and the intellectual freedom to separate what works from what explains

This book is written at that intersection.

It is not an attempt to rewrite history, nor to diminish the achievements that followed Faraday. It is an effort to read him honestly, without imposing later assumptions onto his intuition. To ask what he might have meant, rather than what we later needed him to mean.

In revisiting Faraday now, we are not going backward. We are doing what science periodically requires: *slowing down, re-aligning, and listening again to the original questions.*

Electromagnetism still works flawlessly. Our understanding of it, however, is ready for its next refinement.

This book is written in that spirit.

How to Read This Book

You do not need a background in physics to read this book. You need only one thing: patience with ideas that refuse to rush.

This book is not written to impress, persuade, or overwhelm. It is written to slow explanation down—so that understanding can catch up with familiarity.

What Kind of Book This Is

This is not a textbook. It is not a popular science spectacle. It is not a debate about who was right or wrong.

It is a book about how we explain what we already see working—and how explanation can quietly drift away from honesty when it becomes too confident.

If you have ever felt that science explains *how* things behave but becomes vague when asked *why*, this book was written with you in mind.

How to Approach the Reading

Read slowly, but without fear.

Many chapters build not by adding facts, but by removing unnecessary assumptions. That can feel unusual at first. You may find that nothing dramatic is being introduced—yet something important is being clarified.

If a passage feels calm rather than exciting, that is intentional. Clarity is often quiet.

You do not need to remember terminology. When definitions matter, they are repeated in context. When ideas matter, they reappear from different angles. Understanding comes from recognition, not memorization.

What You Should Not Try to Do

- Do not try to "agree" or "disagree" immediately
- Do not worry about whether ideas are radical or conservative
- Do not try to map everything onto equations or beliefs

Instead, ask one simple question as you read:

Does this explanation make the behavior feel inevitable rather than mysterious?

If the answer is yes, the reading is working.

If You Feel Discomfort

That is normal.

This book removes familiar comfort more than it introduces new ideas. It questions habits of explanation that many of us—scientists and non-scientists alike—have grown used to.

Discomfort here is not confusion. It is often the feeling of letting go of a story that was never required in the first place.

Pause when needed. Reread slowly. There is no penalty for stopping.

What This Book Ultimately Offers

It offers a way of seeing physics—and understanding more broadly—as something grounded in structure, constraint, and necessity, rather than invisible actors and dramatic causes.

You do not need to decide whether you "believe" this framework. Belief is not required. The only thing asked of you is attention.

If you finish this book feeling that the world behaves less mysteriously—but more responsibly—then you have read it well.

A Final Suggestion

Read this book the way its central figure lived his science:

- Observe first
- Explain cautiously
- Stop before inventing

That discipline is not just scientific. It is human.
And it is enough.

A Short Note on UED

What This Framework Assumes (and What It Doesn't)

Before we reinterpret Michael Faraday, it is essential to clarify the lens through which this book views his work. The **Universal Energy Dynamics (UED)** framework is not introduced here as a competing theory of electromagnetism, nor as an alternative mathematical model seeking institutional validation. It is presented as an *ontological clarification tool*—a way of asking cleaner questions about what is actually happening when physical systems change.

UED begins with a modest assumption: *nature prefers equilibrium, and all observable dynamics arise from departures from it.* Nothing in this assumption contradicts classical physics. In fact, conservation laws, stability conditions, and energy minimization principles already reflect this preference implicitly. UED simply makes it explicit and central.

At its core, UED assumes that *energy is not a passive quantity acted upon by forces*, but an active participant that reorganizes itself when configurations change. What we observe as motion, induction, radiation, or coupling is not energy being "pushed" or "pulled," but energy *redistributing to resolve imbalance*. This redistribution occurs locally, through structured interactions, without invoking action at a distance.

Another foundational assumption of UED is that *structure matters*. Geometry, boundaries, orientation, and configuration are not secondary details added after equations; they are primary determinants of how energy couples and evolves. This is why the same components behave differently when rearranged, rotated, or set into motion. Change is not an external trigger—it is the mechanism.

UED also assumes that many entities used in physics—fields, forces, flux lines—are *descriptive constructs*, not necessarily fundamental actors. They are effective languages developed to compress patterns of behavior into workable form. UED does not deny their usefulness. It simply refuses to treat them as ontological commitments unless absolutely necessary. In this sense, UED is conservative rather than radical: it avoids multiplying entities beyond explanatory need.

Just as important as what UED assumes is what it deliberately *does not.*

UED does not claim that existing electromagnetic laws are wrong. Faraday's law, Lenz's law, Maxwell's equations, and all experimentally verified results remain intact. Engineering outcomes remain unchanged. Predictive success is not questioned. What is questioned is the *interpretive story* we tell alongside these results.

UED does not require belief in new particles, hidden dimensions, or exotic forces. It introduces no speculative substances and no unverifiable agents. If a concept cannot be grounded in observable energy behavior and configuration change, UED treats it as a metaphor rather than a mechanism.

UED also does not aim to replace mathematics. Mathematics remains the most powerful language for quantification and prediction. But UED insists that *mathematics follows understanding*, not the other way around. Where equations describe relationships, UED asks what physical reality those relationships reflect. Where symbols succeed, UED asks whether meaning has quietly been outsourced to notation.

Most importantly, UED does not ask the reader to accept conclusions on authority. This framework stands or falls entirely on *coherence, consistency, and explanatory economy*. If a classical explanation is clearer, UED steps aside. If UED clarifies without contradiction, it earns its place.

Within this book, UED is used with restraint. It will not appear as a heavy theoretical overlay, but as a *quiet corrective*—a way of re-reading Faraday's experiments without importing assumptions he never made. The goal is not to turn Faraday into a proto-UED theorist, but to show that his intuition aligns naturally with a framework that prioritizes change, configuration, and equilibrium over invisible actors.

You do not need to "believe" in UED to read this book. You only need to remain open to a simple possibility: that Faraday may have been pointing at something more fundamental than the language we later built around his discoveries.

If that possibility holds, then revisiting his work is not revisionism. It is intellectual housekeeping.

This note is not a declaration of doctrine. It is an invitation to read carefully.

Part I — Before Faraday

Physics Without Change

Phenomena were visible everywhere, yet nothing was moving conceptually.
Nature was observed, cataloged, and measured—but never asked what made it refuse to stay still.

Chapter 1 — Electricity and Magnetism as Curiosities

Static charge, sparks, lodestones, and the absence of unifying logic

A World of Effects Without Explanations

Imagine a room dimly lit by oil lamps. A piece of amber is rubbed with cloth, and suddenly small fragments leap toward it, as if animated by an unseen intent. A spark snaps across a gap—brief, sharp, and gone before it can be examined. Across the room, a compass needle turns obediently, always aligning itself in the same mysterious direction. Elsewhere, iron filings scatter and then arrange themselves into delicate patterns around a dark, unremarkable stone pulled from the earth.

These scenes were not rare. They were repeatable, reliable, and oddly captivating. Long before modern laboratories, people knew how to reproduce them. Rub the amber—attraction appears. Strike the charge—light and shock follow. Bring iron near lodestone—order emerges from chaos. Nature responded consistently, almost politely, to the same gestures.

And yet, beyond that consistency, there was silence.

Electricity appeared as a momentary spectacle—here for an instant, then gone. Magnetism appeared as a permanent oddity—always present, always dependable, yet fundamentally opaque. Both behaved in ways that invited wonder, but neither offered an explanation that could be carried from one observation to another. Each phenomenon stood alone, self-contained, resistant to unification.

What mattered in this early world of inquiry was not *why* something happened, but *that* it happened. Effects were catalogued carefully. Curiosities were demonstrated publicly. Instruments were refined to repeat the same outcomes with greater reliability. But understanding remained fragmented, divided into compartments that never quite spoke to one another.

Electricity was treated as an event—an occurrence triggered by friction or contact, leaving behind no trace of continuity. Magnetism was treated as a property—something inherent to certain materials, useful but inscrutable. The two shared vocabulary—attraction, repulsion, influence—but not meaning. The words overlapped; the ideas did not.

This was not a failure of observation. It was a limitation of interpretation. Nature was performing the same play again and again, but the audience lacked a script. Without a framework to connect these effects, each demonstration became an isolated marvel rather than part of a larger story.

In such a landscape, electricity and magnetism existed as spectacles rather than systems. They impressed, startled, and intrigued—but they did not explain themselves. The phenomena were known intimately, yet their significance remained elusive. Knowledge accumulated, but insight did not.

This was the world into which Faraday would later step—not to add another spectacle, but to ask a question that had been quietly avoided all along: *What changes when these effects appear, and what seeks balance when they do?*

Before that question could be asked, however, the world had to exhaust its fascination with effects alone.

1.1 — Electricity as a Momentary Event

In its earliest encounters, electricity revealed itself only in flashes. A sudden shock to the skin. A sharp spark leaping across a small gap. Lightweight objects drawn briefly toward a charged surface, only to fall away again moments later. These effects were striking, sometimes unsettling, and impossible to ignore—but they were also fleeting.

Electricity appeared, performed its brief spectacle, and vanished.

Because these effects could not be sustained, electricity was naturally treated as something *episodic*. It had no visible persistence. Once the spark faded or the charge dissipated, nothing remained to study. There was no sense of continuity, no lingering presence that could be traced or followed. What could not be held or prolonged could not easily be conceptualized as a system.

As a result, electricity was understood not as a process unfolding over time, but as a *thing that happens* under certain conditions. Rub amber, and attraction occurs. Separate charged bodies, and a spark jumps. Touch the wrong surface, and pain follows. Each event was self-contained, complete in itself, requiring no deeper narrative.

There was also no meaningful idea of *storage or flow*. Without a way to maintain electrical effects, it was difficult to imagine electricity as something that moved, accumulated, or circulated. Motion implies duration. Duration implies structure. Electricity, in this early view, offered neither. It behaved more like a trick of nature than a participant in it.

Even the language reflected this limitation. Electricity was often described as a "fluid" or "effluvia," not because such substances were observed, but because language demanded something tangible to point to. These terms provided comfort, not clarity. They named the effect without explaining its behavior.

Most importantly, electricity had no direction. Sparks jumped where they could. Attractions occurred momentarily and symmetrically. Nothing suggested a preferred path or organized movement. Without direction, there could be no mechanism. Without mechanism, there could be no unifying logic.

In this way, electricity remained trapped in the realm of isolated phenomena. It was dramatic, reproducible, and fascinating—but structurally silent. Each observation reset the story, leaving no trace that could connect one event to the next.

Electricity, as it was then understood, was not something that *operated.* It was something that *happened.*

And as long as it was seen this way, no deeper understanding was possible.

1.2 — Magnetism as a Permanent Oddity

Unlike electricity, magnetism did not arrive in flashes. It was always there. A lodestone pulled iron reliably, without ceremony or effort. A compass needle turned steadily, day after day, pointing in the same direction with unwavering discipline. Magnetism did not surprise; it *persisted.*

This persistence gave magnetism an air of quiet authority. It did not demand attention through spectacle, yet it commanded respect through consistency. Mariners trusted it with their lives. Traders relied on it to cross unfamiliar seas. Practical use flourished long before explanation ever appeared.

And precisely because it worked so well, magnetism was rarely questioned.

Lodestones were treated as natural curiosities—objects endowed with a peculiar ability that set them apart from ordinary matter. Permanent magnets were not seen as participants in a process, but as carriers of a special property. Magnetism belonged to the object itself, much like color or weight. It was something a magnet *had*, not something it *did.*

The compass illustrates this perfectly. Its behavior was indispensable, yet conceptually untouched. The needle aligned itself with no visible guide, responding to something neither seen nor felt. The fact that it always did so reliably made the mystery tolerable. Dependability replaced understanding.

There was no pressing incentive to ask *how* magnetism worked, only reassurance that it did.

Because magnetism did not appear or disappear, it was never framed as an event. Because it did not seem to weaken or exhaust itself, it was not associated with energy. And because it could act across space without contact, it was treated as an intrinsic influence rather than an interaction. The magnet did not respond to its surroundings; it simply expressed its nature.

Language again filled the gap left by mechanism. Magnetism was spoken of in terms of "virtue," "power," or "influence." These words suggested causation without committing to explanation. They allowed magnetism to remain comfortably mysterious while remaining operationally useful.

In contrast to electricity's fleeting displays, magnetism felt grounded and permanent. Yet this very stability concealed its dynamics. What appeared constant discouraged inquiry into what might be changing beneath the surface. The phenomenon was accepted as a fact of nature, not an invitation to investigate structure or process.

Magnetism, in this early understanding, was not something to be explained—it was something to be trusted.

And in that quiet acceptance, an opportunity for deeper insight remained dormant.

1.3 — Two Domains, No Bridge

For centuries, electricity and magnetism occupied neighboring territories in natural philosophy, yet no one built a bridge between them. They were observed, discussed, and demonstrated in the same intellectual circles, sometimes even in the same rooms—but they remained *conceptually isolated.*

Electricity belonged to friction, sparks, shocks, and momentary attractions. Magnetism belonged to lodestones, compass needles, and the quiet certainty of alignment. One was fleeting and dramatic; the other stable and subdued. Their behaviors felt so different that few saw reason to imagine a common foundation.

This separation was reinforced by the absence of any experimental link. Electrical effects could be produced without magnets. Magnetic behavior could be observed without electricity. No obvious action in one domain produced a response in the other. Without such cross-interaction, unification appeared unnecessary, even forced.

Yet the language used to describe both was strikingly similar. Electricity attracted and repelled. Magnetism attracted and repelled. Both appeared to influence objects across empty space. Both acted without visible contact. The same words were used, but they pointed to different ideas. Vocabulary overlapped; understanding did not.

These shared terms gave the illusion of coherence while concealing its absence. Attraction in electricity meant a brief pull following friction. Attraction in magnetism meant a persistent tendency rooted in material. Repulsion in one was a transient event; in the other, a stable orientation. The words matched, but the mechanisms—if they existed at all—were never compared.

As a result, phenomena were *classified rather than unified.* Electricity was given its demonstrations, magnetism its instruments. Each developed its own explanations, metaphors, and experimental traditions. Knowledge expanded sideways, not inward.

This was not due to intellectual negligence. It was a consequence of how science advances when effects precede concepts. Without a reason to connect them, separation was the rational choice. One does not unify what appears unrelated.

But nature, indifferent to human categories, was already doing something subtler. The same underlying order was expressing itself in different guises, waiting for someone to notice not the effects themselves, but *what changed when those effects appeared.*

Until then, electricity and magnetism remained neighbors who never spoke—familiar, useful, and fundamentally misunderstood.

1.4 — The Language Problem

When observation outruns understanding, language rushes in to fill the gap. In the early study of electricity and magnetism, this substitution happened quietly and almost inevitably. Effects were clear, repeatable, and undeniable—but their causes remained inaccessible. To speak about them at all, thinkers reached for the most familiar explanatory tools they had: *substances, fluids, and forces.*

Electricity was described as an *electric fluid*, sometimes single, sometimes dual, flowing in or out of bodies. Magnetism was attributed to a *magnetic virtue*—a hidden quality residing within certain materials. Invisible emissions, or *effluvia*, were imagined to stream from objects, influencing others at a distance. These terms did not arise from evidence; they arose from necessity. Something had to be named, even if nothing had been understood.

This language borrowed heavily from everyday experience. Fluids could flow. Forces could push and pull. Substances could be stored, depleted, or transferred. By adopting these metaphors, early explanations gained a sense of familiarity and comfort. They made the unknown feel manageable. But familiarity is not the same as accuracy.

What these metaphors offered in accessibility, they removed in precision. A fluid that could not be seen, weighed, or contained was not a mechanism—it was a placeholder. A virtue that explained everything explained nothing. These linguistic constructions did not reveal structure; they concealed its absence.

Over time, repetition hardened metaphor into assumption. Once electricity was spoken of as a fluid, it became tempting to reason about it *as if* it behaved like one. Once magnetism was attributed to an inherent virtue, it became unnecessary to ask how it interacted with space or matter. Language began to lead thought rather than serve it.

In this way, explanation drifted toward imagination. Instead of asking how energy might be organized or redistributed, thinkers debated the properties of hypothetical substances. Instead of examining geometry and configuration, they speculated about invisible agents acting behind the scenes. The words multiplied, but insight did not.

This was not intellectual failure; it was an understandable response to conceptual scarcity. Without a clear mechanism, language becomes decorative—ornamenting mystery rather than resolving it. The more compelling the metaphor, the less urgent the need to look beyond it.

By the time Faraday would enter this landscape, electricity and magnetism were burdened not only by unanswered questions, but by *answers that had become too comfortable to question*. Clearing that linguistic fog would prove as important as any experiment.

Only when language loosened its grip could structure finally come into view.

1.5 — Mathematics Without Ontology

As observation accumulated and experiments grew more refined, a natural response followed: *measurement*. If electricity and magnetism could not yet be explained, they could at least be quantified. Distances were measured, angles recorded, strengths compared, and regularities expressed numerically. Mathematics entered the scene as a stabilizing force, offering order where conceptual clarity was lacking.

This development was both necessary and productive. Quantification allowed phenomena to be compared, predictions to be made, and experiments to be repeated with increasing reliability. Numbers replaced narrative. Precision replaced speculation. In practical terms, this was undeniable progress.

Yet something subtle was lost in the transition.

The emerging equations did not arise from a clear picture of what electricity or magnetism *were*. They arose from patterns in what they *did*. Mathematical relationships captured outcomes faithfully, but they remained silent about causes. They told how one quantity changed with another, not why that change occurred in the first place.

This distinction matters.

An equation can predict behavior without revealing mechanism. It can compress observation without clarifying structure. In the absence of ontology—a clear account of what exists and how it interacts—mathematics becomes a powerful descriptive tool, but not an explanatory one.

In this early phase, precision increased rapidly. Constants were refined. Proportionalities were confirmed. Experimental error was reduced. But understanding did not advance at the same pace. The gap between calculation and comprehension quietly widened.

Because the equations worked, they discouraged deeper questioning. Success created confidence, and confidence reduced urgency. If the numbers matched experiment, the explanation felt complete—even if no one could say what, physically, was being described. Mathematics became a substitute for insight rather than its expression.

This imbalance set a precedent that would echo through later physics: the belief that formal correctness could stand in for conceptual clarity. It trained generations to accept calculation as comprehension, and prediction as explanation.

Faraday would instinctively resist this tendency—not by rejecting mathematics, but by refusing to let it lead. He would insist on seeing, visualizing, and re-imagining the physical situation until the behavior made sense without symbols.

This section marks the turning point where that resistance becomes necessary. To move forward, physics would need more than better equations. It would need a new way of thinking—one that returned intuition to its rightful place, not as an enemy of rigor, but as its foundation.

1.6 — Why Curiosity Was Not Enough

By the time the nineteenth century approached, electricity and magnetism had been observed for generations. Demonstrations had become more refined, instruments more sensitive, and descriptions more detailed. Libraries held volumes of recorded effects, classified behaviors, and experimental refinements. By any reasonable measure, curiosity had done its job.

And yet, nothing fundamental had changed.

Despite centuries of attention, electricity and magnetism remained *separate stories*. They were better documented, but no closer to being understood as parts of a single narrative. Each new experiment added another entry to an ever-growing catalogue, but the catalogue itself never transformed into a theory of connection.

This was not because curiosity was absent. On the contrary, fascination was abundant. What was missing was a shift in *how* questions were being asked. Observation had become an end in itself. Effects were admired, repeated, and refined—but rarely challenged at the level of underlying structure.

Progress stalled at description.

Electricity continued to be described in terms of sparks and attractions. Magnetism continued to be described in terms of alignment and pull. Each phenomenon was explained using its own language, metaphors, and mathematical relations. The compartments grew more detailed, but the walls between them grew thicker.

Nature, in this phase, was being watched rather than questioned. The focus remained on *what happens* instead of *what changes when it happens*. Observers stood at a respectful distance, recording behavior without disturbing assumptions. The possibility that the framing itself might be inadequate was rarely entertained.

This limitation reveals an important truth about scientific progress: *curiosity is necessary, but it is not sufficient*. Wonder draws attention, but insight requires interrogation. Seeing is not the same as understanding. Repetition does not guarantee connection.

To unify electricity and magnetism would require more than additional experiments. It would require someone willing to ask a different kind of question—one that cut across categories rather than refining them. A question that treated sparks and needles not as distinct curiosities, but as expressions of a deeper, shared change in nature.

Until such a question was asked, curiosity could only circle the phenomena, never penetrate them.

This is where the story pauses—not because knowledge had ended, but because it had reached the limits of its own approach. To move further, physics needed not more observation, but *a reorientation of attention*.

The stage was set for that shift.

Closing Transition — The Stage Is Set for Change

By the end of this long prelude, the landscape of early electromagnetism was crowded with effects and empty of explanation. Sparks leapt, needles aligned, iron responded, and instruments behaved faithfully—yet no single idea bound these behaviors together. Observation had matured. Language had proliferated. Mathematics had sharpened. And still, logic lagged behind.

The intellectual vacuum was not subtle. Electricity appeared as a sequence of disconnected events. Magnetism persisted as a stable curiosity. Both were reliable, reproducible, and useful. Neither was understood as part of a common process. Effects were everywhere; coherence was nowhere.

This was not because nature was complex beyond reach. It was because attention had been placed on *objects instead of change*, on properties instead of processes, on outcomes instead of transitions. Electricity was treated as something that appeared. Magnetism as something that existed. No one had yet asked what *happened to nature itself* when these effects arose.

What was missing was not data, but orientation.

To move forward, physics did not need a new instrument or a sharper equation. It needed a thinker willing to step back from classification and ask a quieter, more unsettling question: *What changes when these phenomena appear, and what seeks balance when they do?* That question does not begin with things. It begins with relationships.

This is where **Michael Faraday** enters the story—not as a heroic exception, nor as a solitary genius, but as a necessary shift in perspective. Faraday did not arrive to add another phenomenon to the catalogue. He arrived to dissolve the catalogue itself. He did not begin by naming entities. He began by watching how configurations evolved, how influence propagated, and how nature responded to disturbance.

His contribution was not merely experimental. It was philosophical in the deepest sense—though he would never have framed it that way. Faraday instinctively moved attention away from *what things are* toward *what happens between them*. In doing so, he opened a path that had been invisible precisely because it was so simple.

When phenomena refuse to unify, the problem is rarely nature. More often, it is how we have learned to look at it.

With Faraday, that way of looking was about to change.

Chapter 2 — The Comfort Trap of Early Formalism

When equations advanced faster than understanding

Precision Arrives Before Meaning

The era of spectacle gradually gave way to an era of confidence. Where sparks once astonished and lodestones provoked wonder, numbers now stepped in with quiet authority. Measurements replaced marvel. Ratios replaced stories. The unpredictable began to look manageable—not because it was understood, but because it could be counted.

This shift brought relief.

With numbers came predictability. Effects that once appeared temperamental could now be reproduced within known limits. Distances, strengths, and angles were no longer matters of description but of calculation. Experiments could be verified, repeated, and compared across places and generations. Physics began to feel stable, even secure.

Equations offered something deeper than convenience—they offered *control.* To express a phenomenon mathematically was to domesticate it, to bring it within the boundaries of what could be taught, tested, and trusted. Where language had been imprecise, symbols were exact. Where metaphor had drifted, notation held firm.

Authority followed naturally.

A relationship written in symbols carried a weight that no narrative explanation could match. Once behavior could be predicted numerically, the urge to ask what lay beneath it diminished. Precision itself became a form of explanation. If an equation worked, it felt complete.

Yet this new confidence carried a quiet cost.

The move from narrative to number did not clarify meaning; it *replaced* it. Equations described how quantities changed together, but they remained silent on what physically caused those changes. The success of prediction masked the absence of mechanism. Understanding was assumed to be implicit, even when it had never been articulated.

Physics became safer, but not clearer.

The danger was subtle, because nothing appeared to be lost. Results improved. Accuracy increased. Disagreements narrowed. But the central questions—the ones that once provoked wonder—were quietly set aside. As long as calculations matched observation, deeper inquiry felt unnecessary.

This is the paradox at the heart of early formalism. Precision arrived before meaning, and in doing so, it postponed the need to ask what the symbols were truly describing. The language of mathematics grew fluent while the language of understanding fell silent.

It was into this landscape of confidence without clarity that Faraday would later step—not to challenge precision, but to restore meaning to it.

2.1 — Why Formalism Felt Like Progress

Once phenomena could be repeated reliably, quantification became an almost unavoidable next step. Repetition invites comparison, and comparison demands measurement. When sparks jumped the same distance, when needles aligned at consistent angles, when forces produced similar effects again and again, numbers offered a way to bring order to experience. What could be measured could be shared, checked, and trusted.

Measurement reduced ambiguity in ways that language never could. Words vary with interpretation; numbers do not. Two observers might disagree on the intensity of a spark or the strength of a pull, but they could agree on a value recorded by an instrument. In this sense, quantification created a common ground. It narrowed disagreement and accelerated consensus.

Early successes reinforced this trust. Mathematical expressions captured regularities with remarkable efficiency. Relationships that once required lengthy explanation could be compressed into compact formulas. Prediction became possible. Experiments could be designed in advance, outcomes anticipated, errors estimated. For a discipline emerging from centuries of fragmented observation, this felt like a decisive step forward.

Mathematics thus became a stabilizing language. It did not merely describe phenomena; it disciplined inquiry. By insisting on consistency, it filtered out speculation and forced precision. In an environment previously dominated by metaphor and conjecture, this rigor was both refreshing and empowering.

Yet the very strength of formalism concealed its limitation.

The equations worked because they reflected patterns in outcomes, not because they clarified the nature of what was acting. They organized results without explaining why those results occurred. Practical problems were solved even as foundational questions remained unanswered. Motors could be imagined, instruments refined, and predictions improved—while the underlying concepts stayed vague.

This did not feel like a deficiency at the time. On the contrary, success breeds confidence. When a method produces reliable results, its assumptions go unexamined. Formalism felt like progress because it delivered stability, predictability, and authority. It made physics operational long before it made it intelligible.

In this sense, formalism did exactly what it was needed to do. It carried physics forward when understanding lagged behind. But it also planted the seeds of a future imbalance—one in which calculation would be mistaken for comprehension.

Recognizing why formalism felt like progress is essential, not to dismiss it, but to understand why intuition would later need to return—not as an opponent of mathematics, but as its guide.

2.2 — Description Masquerading as Explanation

As mathematical formalism took hold, a subtle shift occurred in how explanation itself was understood. Equations began to capture relationships between measurable quantities with impressive accuracy. When one variable changed, another responded in a predictable way. These relationships could be written, tested, and confirmed. In practice, this was invaluable.

But something important changed beneath the surface.

The equations described *how* quantities were related, not *why* they were related. They mapped correlations faithfully, yet remained silent about mechanisms. Cause was quietly replaced by pattern. As long as one could calculate the outcome, the question of what physically produced that outcome lost urgency.

This shift was rarely explicit. No one declared that explanation was no longer necessary. Instead, explanation gradually came to mean "that which predicts correctly." If an equation worked across experiments, it was treated as sufficient. The distinction between describing behavior and understanding it blurred until it nearly vanished.

Predictive success reinforced this blur. When calculations consistently matched observation, deeper inquiry began to feel redundant. Why ask what was happening behind the symbols when the symbols delivered reliable results? Curiosity was not eliminated; it was *satisfied prematurely*.

Over time, usefulness displaced inquiry. Practical mastery became the benchmark of understanding. The ability to compute replaced the need to conceptualize. The question "What causes this?" was softened into "How does this vary?" and eventually into "Does this give the right answer?"

The phrase *"it works"* acquired an authority of its own. It ended discussions, settled debates, and justified assumptions. Mechanism was deferred indefinitely, not because it was unimportant, but because it seemed unnecessary.

This transformation did not halt progress. On the contrary, it accelerated it. Devices could be improved, experiments refined, and predictions sharpened. Yet progress unfolded within a narrowing conceptual frame. As long as equations succeeded, the underlying reality they described was left uninterrogated.

In this way, description quietly masqueraded as explanation. The language of mathematics became so effective that it concealed its own incompleteness. What was lost was not accuracy, but *depth*.

Understanding did not fail; it was postponed. And with each successful calculation, the postponement felt more justified.

2.3 — Entities Born from Symbols

As equations grew more central, the symbols within them began to acquire a subtle authority of their own. Quantities that were originally introduced as convenient abstractions—ways to summarize patterns in behavior—slowly took on the character of physical existence. What began as notation gradually hardened into *things*.

This transformation was not deliberate. No one set out to turn symbols into substances. It happened through repetition, success, and familiarity. When an abstract quantity appeared consistently in equations that worked, it began to feel real by association. Reliability bred reification.

Language and notation reinforced each other in this process. Once a quantity was named, it invited a noun. Once it had a noun, it invited a role. Fields were spoken of as if they existed independently, forces as if they acted intentionally, fluids as if they flowed through space. Mathematical symbols, paired with everyday language, created a convincing illusion of agency.

Over time, these abstractions solidified into assumed actors. Fields were no longer just a way to describe influence; they became something that *exerted* influence. Forces were no longer shorthand for interaction; they became causes in their own right. Fluids, once metaphorical, were treated as if they possessed substance without substance-like properties.

In this way, ontology emerged accidentally rather than intentionally. Assumptions about what exists were not argued for; they were inherited from notation. The success of equations discouraged scrutiny of the conceptual scaffolding beneath them. If a symbol behaved consistently, it felt justified. If it explained outcomes, it felt sufficient.

This accidental ontology carried consequences. Once entities were assumed to exist, explanations naturally revolved around their behavior rather than the conditions that produced the observed effects. Questions shifted from *what changes in the system* to *what does the entity do*. Attention moved away from configuration and toward imagined actors operating behind the scenes.

Symbols, meant to serve understanding, began to *define it*.

This is a critical turning point. When symbols dictate what is considered real, inquiry narrows to what can be symbolized. Anything that resists easy notation risks being dismissed as vague or unscientific. Intuition, geometry, and process recede—not because they are wrong, but because they are harder to encode.

Faraday would later challenge this drift instinctively. He trusted patterns he could see and manipulate over entities he could not justify. But by the time he entered the conversation, the habit of mistaking symbols for substance was already deeply rooted.

Recognizing this shift is essential, because it explains why later clarification would feel like disruption. Once symbols define reality, questioning them feels like questioning reality itself.

2.4 — The Silent Trade-Off

The rise of formalism brought undeniable gains. Measurements became reliable. Predictions improved. Results could be shared, verified, and reproduced with confidence. Physics grew precise, disciplined, and increasingly powerful. But this progress came with a quiet exchange—one that was rarely acknowledged, and almost never examined.

As formalism strengthened, imagination narrowed.

Only what could be expressed mathematically was granted full legitimacy. Concepts that resisted clean notation—process, configuration, boundary, transition—were treated as secondary, often relegated to informal discussion or philosophical aside. The unspoken rule became clear: if an idea could not be written compactly, it was not yet ready to be taken seriously.

This rule reshaped thinking. Intuition, once the primary guide in interpreting phenomena, became auxiliary. Visual reasoning was tolerated, but not trusted. Conceptual exploration was permitted only as a precursor to equations, never as an endpoint. Understanding was expected to arrive eventually, but only through calculation.

In this environment, explanation was deferred. Rather than clarifying the physical situation first and then expressing it mathematically, the sequence reversed. Equations led; interpretation followed—or failed to follow at all. As long as calculations produced correct results, the absence of intuitive clarity felt acceptable.

This trade-off was silent because it was gradual. No single moment marked the abandonment of imagination. Each step toward formal rigor felt justified, even necessary. Yet collectively, these steps shifted the culture of inquiry. The ability to compute replaced the ability to conceptualize. The confidence of numbers overshadowed the uncertainty of understanding.

What was lost was not correctness, but *freedom*. Freedom to ask questions that had no immediate mathematical answer. Freedom to explore structure without knowing how to quantify it. Freedom to admit uncertainty without losing credibility.

Premature abstraction compressed physics before its foundations were fully understood. It froze certain metaphors into assumptions and elevated notation into ontology. In doing so, it constrained the very intuition that might have guided abstraction more carefully.

This is the cost of the silent trade-off. Formalism delivered reliability, but it also imposed boundaries on thought. It made physics safer, more controllable—but less reflective.

Faraday's contribution would matter precisely because he operated outside this exchange. He would not reject formalism, but he would refuse to let it substitute for understanding. In a landscape shaped by premature abstraction, that refusal would become revolutionary.

2.5 — Authority Through Formalism

As mathematical formalism entrenched itself, it did more than organize knowledge—it began to *govern legitimacy*. The ability to express ideas symbolically became a marker of seriousness, competence, and authority. What could be written in equations was considered rigorous; what could not was treated with suspicion, regardless of insight.

Mathematics quietly assumed the role of an intellectual gatekeeper.

This shift altered who was allowed to speak with authority. Fluency in symbols became a prerequisite for participation in foundational discussions. Conceptual clarity without formal expression was no longer sufficient. An idea might be physically compelling, experimentally grounded, or intuitively coherent—but without the right notation, it struggled to be heard.

Over time, legitimacy became tied less to understanding and more to *symbolic fluency*. The question was no longer "Does this explanation make sense?" but "Can this be written in the accepted language?" When the answer was yes, credibility followed. When it was no, the idea was set aside—not always explicitly rejected, but quietly ignored.

This dynamic reshaped inquiry itself. Questions were increasingly reframed to fit existing notation rather than to probe unexplored structure. Instead of asking what phenomena might be revealing about nature, researchers asked what problems could be solved within the current formal framework. The framework became the filter through which curiosity passed.

Alternative viewpoints did not vanish through debate; they faded through exclusion. Approaches that emphasized geometry, configuration, or qualitative reasoning found fewer venues. Intuition-driven insights were tolerated only when they could be quickly translated into symbols. Those who could not make that translation were seen as incomplete, regardless of the value of their observations.

The result was a subtle inversion. Formalism, originally a tool to serve understanding, became a source of authority independent of clarity. Confidence flowed from notation rather than insight. An explanation could be opaque yet respected, provided it was mathematically sophisticated.

This environment did not suppress progress—it shaped its direction. It rewarded refinement over rethinking, extension over reorientation. It made physics efficient, but it also made it conservative in ways that were hard to detect.

Faraday would later stand outside this structure, not by defying it, but by bypassing it entirely. His authority would come not from symbols, but from consistency between observation and intuition. In a culture where mathematics conferred legitimacy, that stance would be both unusual and transformative.

2.6 — Why This Trap Is Comfortable

The trap of early formalism did not persist because it was imposed. It persisted because it was *comfortable.*

Equations provide a sense of closure that few other tools can match. A problem expressed symbolically feels contained, bounded, and complete. Once written, it appears finished—even when its physical meaning remains uncertain. The mind relaxes in the presence of structure, and mathematics offers structure in its most disciplined form.

This sense of closure is deeply reassuring. Ambiguity feels resolved, not because it has been eliminated, but because it has been *encapsulated.* Unanswered questions are hidden behind symbols that behave predictably. The discomfort of not knowing what is happening physically is softened by the confidence of knowing what to calculate.

In this environment, cognitive comfort quietly replaces conceptual tension.

Conceptual tension is demanding. It requires holding incomplete ideas, tolerating uncertainty, and resisting premature explanation. Formalism removes that burden. It allows inquiry to move forward without resolving foundational questions. As long as results accumulate, unresolved meaning feels like a minor inconvenience rather than a central problem.

This comfort compounds over time. Each successful application reinforces the belief that understanding is progressing, even if the underlying assumptions remain untouched. Progress appears linear and cumulative—new equations build upon old ones, refinements improve precision, and exceptions are absorbed through extension or modification.

Generations inherit this structure already in place. Students learn formulas as finished truths, not as provisional descriptions. The absence of conceptual clarity feels normal because it has always been there. What was once a temporary workaround becomes tradition.

Most importantly, the trap is self-reinforcing. Questioning foundational assumptions feels unnecessary when everything appears to work. Doubt feels inefficient. Re-examination feels like regression. Comfort discourages disruption.

This is why the trap persists across generations. It does not demand belief; it rewards compliance. It does not forbid intuition; it simply makes it unnecessary. And in doing so, it quietly postpones the moment when understanding must catch up with precision.

Faraday's significance lies in the fact that he did not find this comfort persuasive. He remained uneasy where others felt secure. That unease—unfashionable, inefficient, and resistant to closure—would prove essential.

Comfort stabilizes progress. But discomfort is what realigns it.

2.7 — What Was Missed

In the steady advance of early formalism, certain questions never quite surfaced—not because they were unimportant, but because the framework had no natural place for them. Variables were introduced, equations refined, and predictions improved. Yet in this process, *change itself was demoted.*

Change was treated as a parameter to be tracked, not as a cause to be examined. It appeared on the right-hand side of equations as something that varied with time or position, but rarely as the primary driver of physical behavior. The focus remained on quantities *at states*, not on the *transition between states*. What mattered was how much something changed, not what the act of changing meant for the system.

Similarly, configuration—the arrangement, orientation, and geometry of components—was treated as background information. It set the stage but did not participate in the drama. Equations assumed configurations rather than interrogated them. The possibility that rearrangement itself could be causal was largely overlooked. Geometry was acknowledged, but not empowered.

Boundaries suffered the same fate. They were drawn, labeled, and accounted for, but seldom examined as active participants. Surfaces, interfaces, and limits were places where conditions were applied, not where dynamics originated. The idea that boundaries might *govern redistribution* rather than merely constrain it remained implicit, rarely articulated.

Most critically, *energy redistribution*—the heart of all physical change—was left unstated. Conservation laws ensured that energy accounting remained correct, but they did not explain *how* energy reorganized locally when systems were disturbed. The movement of energy from one configuration to another was assumed, not explored.

These omissions were not oversights in the ordinary sense. They were consequences of a framework optimized for calculation rather than comprehension. What could be measured and expressed cleanly was emphasized. What required a shift in perspective was postponed.

As a result, the deepest questions were never written down. Not because they were unanswerable, but because they were *invisible within the prevailing language.* Physics advanced around them, not through them.

It would take a different kind of thinker—one less invested in symbolic fluency and more attentive to physical behavior—to notice what had been quietly set aside. Someone willing to look not at variables, but at *what rearranges when variables change.*

That moment was approaching.

Closing Transition — The Need for a Different Kind of Thinker

By this point, the limitation of early formalism becomes clear—not as a failure, but as an *imbalance.* Mathematics had done what it does best: it stabilized physics, disciplined inquiry, and made prediction reliable. None of this is dismissed here. Without formalism, progress would have stalled long before it accelerated.

But formalism alone could not decide *what deserved to be formalized.*

Equations require guidance. They are powerful only when they are anchored to a clear physical picture. When intuition lags, mathematics does not stop; it continues forward, carrying assumptions silently within its symbols. Precision increases, but meaning thins out. The result is not error, but *disconnection*—a growing distance between calculation and comprehension.

This is where physics needed something it could not derive from symbols alone: a return to intuition disciplined by observation.

Not intuition as guesswork, but intuition as sustained attention to how nature behaves when disturbed. Not rejection of equations, but refusal to let them lead without understanding. What was required was a thinker willing to step outside the comfort of formal closure and remain with the unease of unresolved questions.

That thinker did not need to invent new mathematics. He needed to ask different questions.

Questions not about how quantities relate, but about what *changes* when systems interact. Questions not about entities acting, but about configurations reorganizing. Questions not satisfied by prediction alone, but by coherence across observation.

This is where Michael Faraday enters—not as an opponent of formalism, but as its necessary corrective. Faraday did not reject mathematics because he distrusted it. He set it aside because it was not yet ready to speak honestly about what he was seeing. His loyalty was to behavior first, symbols later.

With Faraday, the direction of inquiry would reverse. Instead of beginning with equations and fitting nature into them, he would begin with observation and allow patterns to suggest structure. Instead of assuming entities and tracking their effects, he would watch how influence spread, how arrangements mattered, and how change propagated through space.

This marks the transition the book now follows: from *equation-first thinking to observation-first understanding*, from symbolic authority to physical intuition, from comfort to clarity.

When symbols lead without meaning, precision becomes a substitute for insight. Faraday's significance lies in refusing that substitution.

The stage is now set for his arrival—not as a historical hero, but as the thinker physics needed at precisely that moment.

Interlude I

When Seeing Is Finally Allowed to Matter

By the end of this part, something important has been established—not a conclusion, but a silence.

Electricity and magnetism were not misunderstood because nature hid itself. They were misunderstood because inquiry learned to feel *complete* too early. Effects were catalogued. Symbols stabilized behavior. Authority settled in. And with that settlement, a certain kind of question quietly disappeared.

The question was not *how accurately can we predict this?* It was *what is actually happening when this changes?*

This part shows a discipline that grew confident before it grew reflective. Formalism delivered reliability, but at the cost of postponing meaning. Language filled gaps where mechanism was unclear. Mathematics conferred authority where understanding was incomplete. Over time, this combination became self-reinforcing. Physics did not stop progressing—but it progressed along tracks laid by its own comfort.

This interlude exists to mark that pause.

It is not an accusation. It is an acknowledgment.

Science does this periodically. It advances through phases where description outruns explanation, where tools mature faster than insight. These phases are not mistakes; they are necessary transitions. But they create a specific kind of intellectual tension—one that can only be resolved when someone is willing to stand slightly out of alignment with the prevailing method.

That person does not need better equations. They need a different posture toward observation.

What physics needed at this moment was not more authority, but permission to look again. Permission to trust what experiments were showing without immediately translating them into inherited language. Permission to remain with ambiguity long enough for structure to reveal itself.

This is where Faraday becomes inevitable.

Faraday did not arrive with a mission to reform physics. He arrived with habits that made reform unavoidable. He did not ask whether an idea fit existing notation. He asked whether it remained faithful to what he could repeatedly observe. He did not seek legitimacy through symbols. He sought coherence through behavior.

His lack of formal mathematical training was not a deficiency to be overcome—it was a constraint that forced clarity. Without the option to hide behind equations, Faraday had to understand phenomena in spatial, relational, and dynamic terms. He had to *see* what others were content to calculate.

This interlude marks the shift from inheritance to insight.

Up to now, the story has been about what physics accumulated and what it quietly set aside. From here onward, the story changes tone. It becomes personal, experimental, and quietly radical. It becomes about a mind that refused to let comfort replace curiosity, and refused to let authority substitute for understanding.

Next part does not present Faraday as a genius in hindsight. It presents him as a necessary corrective—a thinker whose way of seeing restored balance to a discipline leaning too heavily on its own success.

We now move from a world where equations spoke louder than experience to a laboratory where experience insisted on being heard.

Only then does electromagnetism truly begin.

Part II — Michael Faraday: The Intuitive Architect

He did not inherit explanations. He earned attention—patiently, repeatedly, and without permission

Chapter 3 — The Outsider's Advantage

Self-education, experimentation, and intellectual courage

A Mind Trained by Observation, Not Authority

Faraday did not enter science through the front door. He came in from the side—quietly, without credentials, without a chair, without the protective armor of accepted theory. He began not as an academic, but as an apprentice: binding books, reading them carefully, copying passages by hand, and absorbing ideas without being trained to defend them.

This absence of institutional authority is often narrated as a social limitation. In truth, it was a structural advantage.

Faraday did not inherit a framework that needed protecting. He did not arrive with equations that demanded loyalty. He was not burdened by the obligation to reconcile new observations with established notation. What he had instead was a direct relationship with phenomena—unmediated by hierarchy, tradition, or expectation.

Constraint shaped clarity.

Because Faraday lacked formal mathematical training, he could not hide behind symbols. Every explanation had to make sense physically. Every claim had to survive repetition at the bench. If he could not picture a process, he did not trust it. If he could not reproduce an effect, he did not defend it. This discipline was not imposed by ideology; it was imposed by circumstance.

Where others leaned on formalism for legitimacy, Faraday leaned on consistency. Where others debated language, he returned to apparatus. Where authority spoke loudly, he listened longer.

This is the essential difference. Faraday was not disadvantaged by his background; he was freed by it. He learned science without absorbing its comforts. He encountered phenomena before he encountered explanations. As a result, he developed habits that ran counter to the prevailing culture—habits that favored observation over argument, structure over metaphor, and coherence over compliance.

Calling Faraday an outsider misses the point. He was not outside science; he was outside its assumptions.

That position allowed him to see what insiders often missed: that the problem was not the phenomena, but the way they were being framed. His questions did not begin with "Which law applies?" but with "What changes when this happens?" He was not correcting errors; he was bypassing them.

This chapter begins here because Faraday's discoveries did not emerge from sudden brilliance. They emerged from a way of thinking that had been quietly shaped by exclusion from formal comfort. His mind was trained not by authority, but by attention.

Before Faraday changed electromagnetism, he changed the posture of inquiry itself.

3.1 — Learning Without Inheritance

Faraday's education did not arrive packaged as a curriculum. It accumulated slowly, deliberately, and without supervision. He read what he could find, copied what he did not yet understand, and returned to the same ideas repeatedly until they became his own. Books were not authorities to be defended; they were starting points to be tested.

This mode of learning mattered.

Without formal hierarchy, Faraday did not absorb knowledge as doctrine. There was no sequence of concepts he was required to accept before questioning them. He encountered ideas in fragments—experiments here, explanations there—and assembled them through repetition rather than instruction. Understanding emerged from persistence, not progression through a syllabus.

Because he was self-taught, Faraday had no allegiance to prevailing theories. He did not need to reconcile his observations with dominant frameworks, nor did he feel compelled to use accepted language to legitimize his thinking. If an explanation felt incomplete, he did not inherit the reflex to defend it. He simply set it aside.

This freedom shaped his priorities. Faraday trusted *what he could repeatedly observe* more than what he could eloquently explain. Interpretation followed observation, not the other way around. He resisted the temptation to fill gaps with speculation. Where meaning was unclear, he remained patient rather than inventive.

In an era when physics was increasingly defined by formal statements and symbolic authority, this stance was unusual. Faraday learned the behavior of systems before learning how to talk about them. He allowed phenomena to teach him their patterns without forcing them into pre-existing narratives.

The result was a kind of intellectual neutrality that is difficult to cultivate intentionally. Faraday did not set out to challenge assumptions; he simply never absorbed many of them. His understanding was shaped from the ground up, anchored in direct engagement rather than inherited structure.

This is the essence of learning without inheritance. It does not produce faster mastery, but it produces cleaner insight. Faraday's later discoveries would not feel revolutionary to him because they did not overturn beliefs he never fully accepted.

They simply followed from what he had learned to see.

3.2 — Experiments as Conversations, Not Proofs

Faraday did not approach experiments as demonstrations meant to validate a theory. He approached them as questions posed to nature, open-ended and patient. An experiment, for him, was not a final statement but a beginning—a way to see how a system responded when gently disturbed.

This distinction shaped everything he did.

Where others designed experiments to confirm expectations, Faraday designed them to explore behavior. He varied conditions not to isolate a single variable in support of a prediction, but to understand how the whole configuration responded. Small changes in arrangement, orientation, or motion were not nuisances to be controlled away; they were signals worth paying attention to.

Repetition was central to this ethic. Faraday repeated experiments relentlessly, not out of doubt, but out of respect for subtlety. Each repetition was an opportunity to notice something previously overlooked. He did not rush to publish conclusions; he lingered with patterns until they stabilized in his own understanding.

Failure did not discourage him. An experiment that behaved unexpectedly was not a setback—it was an invitation. Where others might dismiss anomalies as error or noise, Faraday paused. He asked whether the anomaly was revealing something the prevailing interpretation could not accommodate. This willingness to remain with irregularity required both patience and humility.

In this sense, Faraday treated observation as a dialogue with nature. He did not interrogate nature aggressively, nor did he impose rigid expectations. He listened. He adjusted. He responded. The apparatus was not merely a tool; it was part of a conversation in which nature answered only the questions it was actually asked.

This ethic stands in contrast to a proof-driven mindset. Proof seeks closure. Conversation seeks understanding. Proof narrows focus. Conversation widens it. Faraday's experiments were structured to keep the conversation open long enough for structure to emerge.

This approach would later allow him to notice connections others missed—not because he was smarter, but because he was less impatient. He was willing to let phenomena teach him how they wished to be understood.

Faraday's greatest discoveries did not arise from decisive moments of confirmation. They emerged from sustained attention to behavior over time. His experimental ethic was not about being right; it was about being faithful to what he observed.

That fidelity would prove transformative.

3.3 — Visual Thinking and Spatial Intuition

Faraday's thinking unfolded in space long before it appeared on paper. Where others reached for symbols, he reached for sketches. Diagrams, drawings, and physical arrangements were not auxiliary aids to his reasoning; they were the reasoning itself. To understand a phenomenon, Faraday needed to see how it occupied space.

This preference was not stylistic. It was methodological.

Without formal algebra to lean on, Faraday cultivated a powerful form of spatial intuition. He arranged apparatus, observed how effects propagated, and mentally traced the paths of influence through air, matter, and boundary. Geometry mattered more to him than numerical magnitude. Orientation mattered more than scale. Configuration mattered more than quantity.

The idea of "lines of force" emerged from this habit. These lines were not entities for Faraday. They were visualization tools—ways to represent how influence was distributed through space when a system was disturbed. Iron filings around a magnet did not reveal threads in the air; they revealed structure in response. The pattern mattered, not the metaphor.

Where others saw variables changing with position, Faraday saw patterns reorganizing themselves. A shift in arrangement could transform behavior without altering material. Rotation, proximity, alignment—these geometric factors carried explanatory weight. He noticed that the same components behaved differently when their spatial relationship changed, even if all measurable quantities remained similar.

This way of seeing allowed Faraday to grasp connections that algebra alone could not easily reveal. He was not tracking isolated quantities; he was tracking configurations evolving. Cause, for him, was not located in a variable but in the arrangement of the system as a whole.

In modern terms, Faraday was thinking structurally. He was attentive to how influence filled space, how boundaries shaped response, and how changes propagated through configuration rather than through objects acting at a distance. These insights did not require equations to be valid. They required clarity of perception.

Faraday's reliance on spatial intuition was not a limitation to be overcome by later formalism. It was a strength that formalism would later need to recover. His mind operated in the domain where structure precedes symbol.

He did not ask what equation described the system. He asked what the system *looked like when it changed.*

3.4 — Intellectual Courage Without Confrontation

Faraday did not challenge authority by opposing it. He challenged it by outlasting it.

Faraday's disagreements were rarely public and never theatrical. He did not argue against prevailing views in journals or lecture halls. He did not position himself as a contrarian. Instead, he allowed his experiments to accumulate quietly, returning to the bench again and again until patterns became undeniable. Where others debated, he repeated. Where others defended language, he refined observation.

This restraint is often mistaken for timidity. It was anything but.

Faraday's courage lay in his willingness to disagree silently—to continue trusting what he could reproduce even when it did not align with accepted explanation. He did not seek validation through confrontation. He sought clarity through consistency. If an idea survived repeated experimentation, he allowed it to stand, regardless of how unfamiliar it sounded.

There were no authority battles because Faraday did not measure truth by reputation. He did not need to win arguments to be confident. His assurance came from repeatability. If a result held across configurations, variations, and time, it earned his trust. If it did not, he let it go—no matter how elegant it appeared on paper.

This approach demanded patience. It also demanded humility. Letting experiments speak means accepting that nature may contradict expectation, language, or prestige. It means allowing evidence to mature before interpretation. Faraday was willing to wait.

In a culture increasingly shaped by formal authority and symbolic fluency, this posture was quietly radical. Faraday refused to escalate disagreement into conflict. He did not attack theories; he rendered them unnecessary by showing something more coherent.

This is a different kind of intellectual courage—one that does not seek to overthrow, but to persist. One that trusts process over proclamation. One that understands that clarity, once achieved, does not need volume.

Faraday's confidence was not loud. It was durable.

And in the long run, durability proved more disruptive than defiance.

3.5 — Freedom from Formal Comfort

Faraday's inability to rely on mathematics was not a temporary obstacle on the way to mastery. It was a permanent condition—and a decisive one. Without the option to retreat into symbols, he was forced to confront every idea at the level of physical sense-making. If a concept could not be visualized, arranged, or tested through configuration, it did not earn his confidence.

This absence of formal comfort imposed a discipline that mathematics often postpones.

Faraday could not declare a result "explained" because an equation balanced. He had to resolve his intuition in the laboratory, not on the page. Every explanation had to survive contact with apparatus, geometry, and repetition. Where symbols can compress uncertainty, Faraday had to live with it until clarity emerged.

This constraint shaped his standards. He refused to accept explanations he could not picture spatially or reproduce experimentally. If a description relied too heavily on invisible actors or abstract forces without observable structure, he treated it with suspicion. Not rejection—suspicion. He simply did not build on what he could not *see working*.

What emerged instead was a form of precision without symbols. Faraday's work was meticulous, controlled, and repeatable. His experiments were not vague explorations; they were carefully structured investigations of how systems responded to change. Precision was achieved through arrangement, timing, orientation, and boundary—not through notation.

This is an important distinction. Precision does not belong exclusively to mathematics. It belongs to consistency of behavior. Faraday achieved it by controlling what he could physically manipulate and observing how outcomes shifted when configurations changed. His rigor lived in space and sequence rather than in formulae.

The key insight here is simple but profound: constraint preserved Faraday's honesty.

Without mathematical shortcuts, he could not mistake calculation for comprehension. Without symbolic authority, he could not confuse elegance with truth. Every claim had to earn its place through coherence with observation. This made his thinking slower—but also cleaner.

Faraday did not reject mathematics. He simply refused to let it substitute for understanding. In doing so, he protected his intuition from becoming decorative or complacent.

That freedom—from formal comfort—would later allow him to see what others, surrounded by symbols, had learned not to look for.

3.6 — Why This Mindset Mattered

Faraday's way of thinking mattered not because it was different, but because it arrived *at exactly the point where physics had become too comfortable with its own tools.* While many of his contemporaries worked equation-first—beginning with formal relationships and fitting observations into them—Faraday worked in the opposite direction. He began with behavior and allowed structure to emerge before language followed.

This contrast is essential.

Equation-first thinkers were skilled at refining what was already visible within existing frameworks. They improved precision, extended reach, and increased reliability. But their attention was naturally constrained by the symbols they used. What could not be expressed cleanly was often postponed. What did not fit the form was treated as secondary.

Faraday, working without that symbolic safety net, had no such constraint. Because he was not trying to preserve a formal structure, he was free to notice connections that crossed categories. Electricity and magnetism did not appear to him as separate domains needing reconciliation; they appeared as different expressions of how systems responded when disturbed. He was not looking for a bridge between theories. He was watching how change propagated.

This is why Faraday could see what others missed. He was attentive to transitions rather than states, to rearrangements rather than properties. When a current began or ended, when a magnet moved or rotated, when a configuration shifted—even slightly—he noticed that the system behaved differently. These differences were not anomalies to be averaged out; they were signals.

Observation-first thinking prepared the ground for induction long before induction was named.

Faraday was already asking the right kind of question: not *what force is acting?* but *what changes when this changes?* That orientation made it possible for him to recognize that motion, switching, and variation were not incidental triggers—they were central causes. Induction did not arrive as a surprise; it arrived as a natural consequence of sustained attention to change.

From the perspective of the Universal Energy Dynamics (UED) framework, this mindset appears strikingly aligned. Faraday's emphasis on configuration, boundary, and response anticipates a view in which energy redistribution seeks equilibrium whenever a system is disturbed. He did not articulate this formally, but he worked as if it were true.

The seeds were already there:

- Change as primary
- Configuration as causal
- Geometry as explanatory
- Stability as a guiding principle

Faraday did not need a unifying equation to recognize unity. He recognized it through consistency of behavior across different situations. That recognition is what allowed induction, motors, and later field concepts to emerge—not as inventions, but as clarifications.

This section completes the preparation.

What follows in the next chapters is not a sequence of accidental discoveries, but the natural unfolding of a mindset that had been quietly trained to see what formalism alone could not. Faraday's breakthroughs did not break physics. They realigned it.

The work of intuition is now ready to become visible.

Closing Transition — From Mindset to Motion

By the time Faraday approached his first major discovery, nothing about his method was accidental. The habits that shaped his thinking—learning without inheritance, experimenting without haste, trusting spatial intuition, and refusing formal comfort—had already aligned his attention in a particular direction. He was no longer looking for laws to apply. He was watching what systems did when they were disturbed.

This distinction matters.

Faraday did not wake up intending to invent a motor. He was not chasing application or utility. He was exploring a question that had quietly guided his work: *How does electrical activity alter the state of a system when configuration changes?* Motion was not his goal. It was the system's response.

When electrical current was introduced into a physical arrangement involving magnets, Faraday noticed something that others had overlooked—not because it was hidden, but because it did not fit prevailing expectations. The system did not settle. It reorganized continuously. Equilibrium was never reached. Instead, motion persisted.

That persistence was the clue.

The electric motor of 1821 was not a clever trick assembled from known parts. It was the visible outcome of a mindset trained to notice sustained non-equilibrium. Faraday did not interpret motion as a force acting on an object. He saw it as a configuration that could not resolve itself into rest. Change fed change.

This is where method reveals its power. Had Faraday approached the problem equation-first, motion might have appeared as an anomaly or an artifact. Had he been committed to inherited explanations, he might have dismissed it as incidental. Instead, he trusted what he observed and followed it.

Discovery followed method, not chance.

The transition from mindset to motion marks a turning point in this book. Up to now, we have traced how Faraday learned to look. From here onward, we will see what that way of looking made visible. The laboratory will replace reflection, and intuition will take material form.

Before Faraday changed electromagnetism, he changed how it was looked at.

With that change in place, motion was inevitable.

Chapter 4 — Motion from Electricity (1821)

The first electric motor and the discovery of sustained non-equilibrium

When Stillness Refused to Return

In 1821, the setup could not have looked more ordinary. A wire, a source of current, a magnet—nothing exotic, nothing ornate. The components were familiar, even mundane, to anyone experimenting with electricity at the time. There was no elaborate machinery, no hidden mechanism waiting to be revealed. The arrangement appeared simple enough to predict.

The expectation was equally simple.

Introduce a current, observe a response, and watch the system settle back into rest. That was how electrical effects behaved. They appeared briefly, did their work, and vanished. Motion, if it occurred at all, was expected to be momentary—a deflection, a twitch, a transient adjustment before equilibrium reclaimed the system.

That is not what happened.

When Faraday closed the circuit, the wire did not merely move. It rotated. And it did not stop. As long as the current flowed, motion persisted—smooth, continuous, and unmistakably sustained. Stillness did not return.

This persistence was the shock.

Nothing in the prevailing understanding of electricity prepared observers for continuous motion arising so effortlessly. There was no obvious push, no single direction of force being applied. No part of the system appeared to be "driving" another in the familiar mechanical sense. Yet motion endured, refusing to decay into rest.

What Faraday encountered in that moment was not an invention seeking application, but a surprise demanding interpretation. The apparatus had not been designed to produce a motor. It had been arranged to observe interaction. Motion emerged not because it was engineered, but because the system could not resolve itself into equilibrium.

This violated a deeply held expectation: that disturbances naturally die out. In most physical systems, imbalance leads to adjustment, and adjustment leads to rest. Here, imbalance led to continuity. Change fed change.

The significance of this moment lies not in the rotation itself, but in what it implied. Motion was no longer an event triggered and exhausted. It was a state sustained by configuration. The arrangement of components mattered more than their individual properties. The system did not seek stillness; it organized around motion.

Faraday did not celebrate this as triumph. He did not rush to name it. He watched it, repeated it, and let its refusal to stop speak for itself. What stood before him was not a clever trick, but the first visible sign that electricity could participate in ongoing physical processes rather than isolated effects.

Something fundamentally new had appeared—not a force pushing an object, but a configuration that could not settle.

Electromagnetism had taken its first step out of stillness.

4.1 — The Experiment That Should Have Stopped

Faraday's 1821 apparatus was almost disarmingly simple. A current-carrying wire was allowed to move freely near a magnet, arranged so that electrical contact could be maintained while motion was possible. There were no gears, no coils wound for torque, no mechanical ambition built into the design. It was an arrangement meant to observe interaction, not to produce sustained work.

Given everything known at the time, the expectation was straightforward.

When current was introduced, the wire might shift position slightly—perhaps deflect, perhaps realign—responding briefly to whatever influence existed between electricity and magnetism. Then, having adjusted, it should come to rest. This was how physical systems behaved. Disturbance led to response; response led to equilibrium. The story was always short.

Instead, the story refused to end.

As soon as the circuit was completed, the wire began to move—and then continued to move. It did not overshoot and settle. It did not oscillate and dampen. It rotated, smoothly and persistently, as long as the current was maintained. Remove the current, and motion ceased. Restore it, and motion returned. The behavior was not erratic; it was disciplined.

This was deeply unsettling.

Nothing in the apparatus suggested a stored source of mechanical energy. There was no obvious push being applied in a fixed direction. No single moment of cause could be isolated. The motion was not being "spent"; it was being sustained. The system behaved as if rest were no longer the preferred outcome.

For a physics culture trained to expect equilibrium, this was a quiet rupture. Motion had always been understood as something that required continual external forcing or internal energy storage. Here, motion arose from configuration alone—from the way current, magnet, and freedom of movement were arranged together.

The experiment should have stopped. Nature declined.

This refusal to settle back into equilibrium was not a small anomaly to be corrected. It was a signal that the prevailing intuition about electrical effects was incomplete. Electricity was no longer behaving like a momentary event. It was participating in an ongoing process.

Faraday did not rush to explain this behavior. He did not dress it in new language or force it into old categories. He simply allowed the persistence of motion to stand as a fact—repeatable, controllable, and undeniable.

The key insight was not that electricity could cause motion. That much could have been imagined. The insight was that *motion could be the stable outcome*, not the transient one. Change did not dissipate; it organized.

Nature, in this experiment, refused to return to rest.

And in doing so, it quietly announced that equilibrium was no longer the whole story.

4.2 — Why This Was Not "Force at Work"

The most natural way to describe motion is to imagine a push. Something applies a force, something responds, and movement follows. This intuition is deeply ingrained, shaped by everyday experience and reinforced by centuries of mechanical reasoning. When something moves, we look for what pushed it.

Faraday's rotating wire resisted that instinct.

There was no single direction of applied force acting on the wire. Nothing struck it, pulled it, or continuously pushed it along a fixed line. If one tried to locate a point where a force was being "applied," it dissolved under inspection. The motion did not originate at a contact point. It did not proceed along a straight path dictated by a push. It curved, adapted, and continued.

This is where force-centric language begins to strain.

In a push–pull picture, force has direction, location, and magnitude. It acts here, produces motion there, and can be traced from cause to effect. Faraday's experiment offered none of these conveniences. The motion was not aligned with a single direction of action. It changed orientation continuously as the wire moved. Any attempt to assign a fixed force vector felt artificial—imposed rather than revealed.

What *did* remain consistent was the geometry of the system.

The relative arrangement of the current, the magnet, and the freedom of movement mattered far more than any imagined push. Alter the configuration, and the behavior changed. Preserve it, and motion persisted. The cause was not localized; it was distributed across the arrangement as a whole.

This is why force language struggles here. Forces are well-suited to describe interactions between objects. They are poorly suited to describe behaviors that arise from system-wide asymmetry. In Faraday's setup, nothing acted on the wire in isolation. The wire was participating in a configuration that could not resolve itself into rest.

Motion emerged not because something was being driven, but because there was no path to stillness.

To say "a force caused the motion" is not wrong in a narrow sense, but it is incomplete in a crucial one. It shifts attention away from what actually mattered—the arrangement—and toward an imagined agent. It names an effect without clarifying the mechanism.

Faraday sensed this intuitively. He did not insist on describing the motion as the result of a force acting on a current. He focused instead on how the interaction filled space, how orientation mattered, and how the system behaved as a whole. He resisted the temptation to reduce a distributed phenomenon to a localized cause.

This resistance was not philosophical. It was practical. Force language did not help him understand what he was seeing.

The motion was real. The push was not.

By allowing the experiment to challenge force-centric intuition rather than forcing it to conform, Faraday opened the door to a deeper interpretation—one in which configuration, not force, became the primary explanatory element.

That shift would soon prove decisive.

4.3 — Configuration as the Cause

What distinguished Faraday's 1821 experiment was not the novelty of its components, but the specific way they were arranged. A conducting wire, a magnet, and a point of support—each was familiar on its own. None possessed any mysterious property capable of producing continuous motion in isolation. Yet when brought together in a particular spatial relationship, the system behaved in a way that no single component could explain.

This is the first moment where configuration asserts itself as causal.

The wire was free to move, but constrained to a geometry. The magnet was fixed, but oriented in space. The current flowed, but not along a straight mechanical path. These elements formed a relational structure in which every part mattered—not because of what it was, but because of where it was and how it was oriented.

Change the geometry, and the behavior changed. Reposition the magnet, and the direction of rotation altered. Modify the point of contact, and the motion weakened or disappeared.

Nothing about the material properties had changed. Only the arrangement had.

This observation quietly undermined a deeply held assumption: that causes reside in objects. In Faraday's setup, no object "contained" the cause of motion. The cause was distributed across the spatial configuration. Motion was not being produced by a component; it was emerging from the relationship between components.

The key to this emergence was imbalance—but not the kind that could be easily resolved. In most systems, imbalance leads to adjustment and rest. Here, adjustment simply recreated imbalance in a new orientation. The system had no stable resting state available within its constraints. As long as current flowed, any attempt to settle only redirected the system back into motion.

This is what makes the imbalance unresolvable.

The system was not being driven forward by a force; it was being denied equilibrium by its own geometry. Motion was the only behavior compatible with the configuration. Rotation was not a response to an external push—it was the system's way of continuously negotiating an arrangement that could not stabilize.

In recognizing this, Faraday caught an early glimpse of a principle that would later reshape electromagnetism: dynamics can arise from configuration alone. The arrangement of elements can be more important than the elements themselves. Geometry can act where force language fails.

This was a quiet but profound shift. It suggested that motion, energy transfer, and sustained activity might be explained not by isolated causes, but by how systems are put together. It hinted that understanding electromagnetism would require attention to structure, orientation, and boundary—not just to quantities and magnitudes.

The arrangement produced motion. The components merely participated.

With this realization, electromagnetism began to move away from objects and toward relationships—a direction Faraday would follow with increasing confidence in the years to come.

4.4 — Sustained Non-Equilibrium

Most physical systems share a quiet preference: they want to settle. Disturb them, and they respond. Let them respond long enough, and they relax back into stillness. A stretched spring returns to rest. A displaced object oscillates and eventually stops. Even dramatic disturbances tend to fade as balance is restored. This tendency feels so natural that it is often taken for granted.

This is what equilibrium looks like in everyday experience.

Non-equilibrium, by contrast, is simply a state in which that settling does not occur. The system remains active, not because it is being violently driven, but because its conditions prevent it from resting. Something about its arrangement keeps the imbalance alive.

In most cases, non-equilibrium is temporary. Energy enters, rearrangement happens, and dissipation smooths everything out. Rest returns. That expectation was deeply ingrained in how electricity was understood as well—effects appeared, did their work, and vanished.

Faraday's rotating system broke that expectation cleanly.

Here, the system did not relax. It did not oscillate toward rest. It did not decay. As long as current flowed, motion persisted. Remove the current, and stillness returned instantly. Restore it, and motion resumed just as reliably. The behavior was stable, repeatable, and controlled—but never settled.

Why could this system not come to rest?

Because any attempt to do so recreated the very imbalance that initiated motion. The configuration offered no position of stability. Adjustment did not resolve the situation; it reproduced it in a new orientation. The system was caught in a loop where change was the only available response to imbalance.

This is the key insight: the motion was not a sign of instability or loss of control. It was the system's *best possible response* to its constraints. Continuous rotation was not excess behavior—it was the only behavior compatible with the arrangement.

Seen this way, the motion represents an equilibrium-seeking failure. The system continually attempts to stabilize, but the geometry prevents stabilization from ever completing. Motion is not imposed on the system; it is what equilibrium-seeking looks like when equilibrium is unreachable.

This reframes how activity itself is understood. Motion no longer needs to be explained as something caused and maintained by force. It can be understood as the natural outcome of a configuration that cannot resolve its internal imbalance.

Faraday did not name this concept. He did not formalize it. But he recognized it in practice. He saw that motion could be sustained without escalation, persistent without chaos, and controlled without exhaustion.

This recognition marks a turning point.

Electricity was no longer just something that disturbed systems briefly. It had revealed its capacity to participate in ongoing dynamics—to keep a system active not by pushing it forward, but by preventing it from stopping.

That realization would soon echo far beyond motion itself.

4.5 — Why Others Missed It

In hindsight, Faraday's 1821 experiment appears almost obvious. The apparatus was simple. The motion was visible. The result was repeatable. Yet for many of his contemporaries, the significance of what he had observed passed quietly by. The question is not why they failed to see the motion, but why they failed to see what the motion meant.

The answer lies in expectations.

Most physicists of the time approached phenomena equation-first. They carried with them an implicit belief that behavior should conform to known forms. Motion, in particular, was expected to be resolvable into forces acting in identifiable directions. When an observation did not naturally fit this picture, it was treated as an oddity rather than a doorway.

Faraday's setup looked trivial because it did not announce its importance through complexity. There was no elaborate instrumentation to signal depth. No new quantity to measure. No obvious extension of existing theory. To an equation-trained eye, it looked like a laboratory curiosity—interesting, perhaps, but conceptually thin.

This is where Faraday differed.

He did not ask whether the behavior could be immediately explained. He asked whether it persisted. He trusted duration more than description. The fact that motion continued as long as current flowed mattered more to him than whether he could name the cause. Persistence was the signal. Explanation could wait.

Others were inclined to move on once they had classified the effect. The wire moves. The magnet interacts. End of story. Motion was treated as a demonstration, not a principle. It was something to be shown, not something to be dwelled upon.

But Faraday dwelled.

He stayed with the instability that others found uncomfortable. Sustained non-equilibrium challenges a deeply ingrained intuition—that systems should calm down. For many, that challenge was easier to ignore than to investigate. Instability felt like noise rather than structure.

Faraday had learned, through years of patient experimentation, that instability could be informative. A system that refuses to settle is telling you something about its constraints. Ignoring that refusal means missing the message.

The key insight is subtle but decisive: seeing sustained change requires patience with instability. It requires resisting the urge to normalize behavior too quickly. It requires allowing a system to remain unsettled long enough for its logic to reveal itself.

Faraday possessed that patience. His contemporaries, trained to value closure and elegance, often did not.

As a result, what looked like a modest laboratory effect to many became, in Faraday's hands, the first clear sign that electricity could sustain ongoing dynamics. He did not see a curiosity. He saw a principle waiting to be understood.

And once seen, it could not be unseen.

4.6 — Reinterpreting the Motor Under UED

Viewed through the Universal Energy Dynamics (UED) lens, Faraday's motor stops looking like a clever mechanical arrangement and starts revealing itself as something more fundamental: a dynamic equilibrium system. The rotation is not an added feature; it is the system's natural way of managing imbalance.

In UED terms, motion is best understood as ongoing energy redistribution. Energy does not act once and disappear. When introduced into a system, it seeks equilibrium by reorganizing how it is distributed across space, boundaries, and configurations. In Faraday's experiment, electrical current injects energy continuously, not as a single impulse but as a sustained condition.

That current introduces asymmetry.

The system is no longer neutral. The presence of current breaks spatial symmetry, creating a directional bias in how energy can redistribute. Importantly, this asymmetry is not localized at a point; it is spread across the configuration. The wire, the magnet, and the surrounding space participate together in shaping how redistribution unfolds.

Under ordinary circumstances, asymmetry resolves. Energy flows, adjustments occur, and equilibrium is restored. But in Faraday's motor, geometry prevents resolution. The spatial arrangement ensures that any attempt to settle recreates imbalance in a new orientation. The system is perpetually "almost balanced" but never fully so.

This is the decisive insight.

Because equilibrium is unreachable, the system adopts the only response that remains stable under continuous energy input: rotation. Rotation is not excess behavior; it is the lowest-energy pathway available given the constraints. The system is not fighting equilibrium—it is seeking it and failing gracefully.

From a UED perspective, this reframes the motor completely. There is no need to imagine invisible forces pushing the wire around. The motion is not driven by a hidden agent. It is the visible signature of a system managing energy flow under constrained geometry. Continuous motion is simply what equilibrium-seeking looks like when equilibrium cannot be achieved.

This is why the motor is not a trick.

A trick exploits mechanics to create motion. Faraday's motor reveals a principle: when energy input, asymmetry, and configuration align in a particular way, motion becomes the stable state. Stop the current, and redistribution halts. Change the geometry, and the pathway collapses. Preserve both, and rotation persists.

Seen this way, the motor is an early demonstration of a core UED idea: dynamic systems do not always settle into stillness; sometimes they settle into activity.

Faraday did not articulate this framework, but his experiment embodied it perfectly. He showed—without equations—that sustained motion can arise naturally from how energy, space, and configuration interact.

Electromagnetism, through this lens, is no longer about objects acting on each other. It is about systems reorganizing themselves under continuous imbalance.

And that realization would soon make the next step—induction—not only possible, but inevitable.

4.7 — From Curiosity to Consequence

At first glance, Faraday's rotating wire could have remained a laboratory curiosity—an elegant demonstration to be shown and then set aside. Many such effects existed already, briefly entertaining and quickly forgotten. What distinguished this experiment was not the novelty of motion, but the consequence of its persistence.

The moment motion refused to stop, electricity crossed a conceptual threshold.

Until this point, electrical phenomena were largely treated as events. A spark jumped. A shock occurred. A charge accumulated and discharged. Even when impressive, these effects were episodic. They appeared, did something observable, and disappeared. Electricity was dramatic, but fleeting.

Faraday's experiment changed that narrative.

Here was electricity participating in continuous work. Motion was not a side effect; it was sustained as long as electrical conditions were maintained. This meant that electricity was no longer just something that happened—it was something that could *operate*. It could maintain activity, not merely trigger it.

This reframing was profound.

Once electricity could sustain motion, it could no longer be understood as an isolated phenomenon. It had to be understood as a process unfolding over time. Attention shifted from what electricity does at a moment to what it does as it flows. Duration became as important as intensity. Configuration became as important as magnitude.

Within this shift lay the seeds of induction.

If electricity could maintain motion through continuous redistribution, then change itself—starting, stopping, moving, rotating—deserved closer scrutiny. The question naturally evolved: *What happens in the surrounding system when electrical conditions change?* Faraday's motor did not answer that question, but it made it unavoidable.

This is how curiosity matured into consequence.

Faraday did not immediately leap to generalization. He did not claim a new law. But the experiment quietly rewired his thinking. Electricity was no longer an episodic disturbance; it was a participant in sustained dynamics. It behaved more like a flow than a flash, more like a process than an event.

That realization would guide his attention over the next decade. It would make him sensitive to transitions—beginnings, endings, motion, variation. It would lead him to ask what occurs not just where electricity is present, but where it changes.

In this sense, the motor was a preparation. It trained Faraday's intuition to think in terms of continuity, redistribution, and configuration. It expanded the role of electricity from spectacle to system.

Electricity had become operational.

And once electricity was understood as a process rather than an event, the door to induction was already open—waiting for Faraday to step through it.

Closing Transition — Motion Points Forward

Faraday did not linger on motion as an endpoint. The rotating wire was not a destination; it was a signal. Once he recognized that electricity could sustain activity rather than merely trigger it, a deeper question emerged—one that motion alone could not answer.

If continuous motion was possible, then change itself had agency.

The motor showed that a system could remain active as long as conditions were maintained. But Faraday's attention shifted naturally to what happened *when those conditions changed.* When current began or ended. When orientation shifted. When motion was introduced rather than sustained. He began to suspect that the most important effects were not located in steady states, but in transitions.

This was a decisive turn.

Motion had revealed that electricity was not episodic. Now Faraday wondered whether *variation*—starting, stopping, moving, rotating—might produce effects of its own. What happens not just where electricity is present, but where it changes? What happens in the surrounding system when conditions are altered, even briefly?

The motor made this question unavoidable.

It suggested that systems do not merely respond to electricity; they respond to reorganization. That response might not require contact. It might not require sustained current. It might arise from change propagating through space itself.

Faraday did not yet know the answer. But his method was already aligned to find it.

He would spend the next decade following this intuition patiently—varying conditions, interrupting currents, moving magnets, rotating conductors—watching how systems responded not to presence, but to difference. The culmination of that attention would arrive in 1831, when induction finally made the role of change explicit.

This chapter ends where the next one must begin.

Motion proved that electricity could operate continuously. Induction would prove that change itself could act.

Once motion could be sustained by change, electricity could no longer be treated as an event. It had become a process—one that was about to reveal its deepest principle.

Next chapter awaits.

Chapter 5 — Induction (1831)

When Change Became the Cause. Why motion, switching, and rotation all induce current

A Decade Later, a Different Question

Ten years had passed since motion first refused to stop.

In that decade, Faraday did not chase applause, nor did he rush to generalize what he had seen in 1821. He returned to the laboratory with a quieter, sharper question—one that motion itself had planted but not answered.

The motor had shown *what* electricity could do. Now Faraday wanted to understand *when* it did anything at all.

He noticed something unsettling in its simplicity. A current flowing steadily produced no surprises. A magnet resting quietly near a conductor did nothing remarkable. Presence alone—no matter how intense—seemed strangely ineffective. The system remained silent unless something *changed*.

This realization reframed everything.

Faraday began to suspect that nature was indifferent to static conditions. What mattered was not what existed, but what varied. Not where electricity was, but where it began, ended, moved, or reoriented. The cause of effects was slipping away from objects and settling into transitions.

This was not a dramatic revelation. It did not arrive as a single experiment or a sudden insight. It emerged slowly, through repetition and restraint. Faraday resisted the temptation to declare a principle before it proved itself across situations. Instead, he waited—testing different arrangements, interrupting currents, moving magnets, rotating conductors—watching closely for the common thread.

What he found was unsettling in its consistency.

Whenever nothing changed, nothing happened. Whenever something changed, something responded.

It did not matter how the change occurred. Switching a current on or off. Sliding a magnet closer or farther. Rotating a loop continuously. Different actions, same consequence. Current appeared where none existed before—not because something acted upon it, but because the system could not remain indifferent to change.

By 1831, Faraday was no longer studying electricity or magnetism as separate phenomena. He was studying change itself—how systems respond when equilibrium is disturbed, even momentarily. Induction was not discovered as a law; it revealed itself as a pattern that refused to go away.

This opening scene matters because it marks a decisive shift in causality.

Before induction, causes were sought in things—forces, charges, magnets. After induction, the cause was unmistakably elsewhere. It lived in variation, in transitions, in the act of becoming rather than the state of being.

Faraday did not yet have the language to formalize this shift. He did not need it. The experiments were already speaking clearly.

Induction did not say, *"Objects act."* It said, *"Change demands response."*

With that realization, electromagnetism crossed a threshold it could never retreat from.

5.1 — The Failure of Presence

Faraday's first realization in the long path to induction was deceptively simple—and deeply disruptive. He discovered that presence alone did nothing.

A current flowing steadily through a wire produced no new effect in a nearby conductor. A magnet resting quietly beside a loop induced nothing. Even strong arrangements, impressive in their stillness, remained inert. Nothing appeared, nothing transferred, nothing responded. The system was calm—almost indifferent.

This was not what prevailing intuition suggested.

If electricity or magnetism were active agents, then their mere presence should have mattered. If a magnet *acted*, then being nearby should have been enough. If a current *did something*, then sustaining it should have produced a sustained effect elsewhere. Yet experiment after experiment contradicted this expectation.

Faraday noticed a pattern that others had overlooked because it was too quiet to be dramatic.

When everything was steady, nothing happened.

This was unsettling because it challenged the very idea of static causality. The assumption that "something exists, therefore something acts" failed repeatedly at the bench. No matter how carefully Faraday arranged the apparatus, no matter how intense the current or how strong the magnet, stillness produced silence.

What failed here was not technique, but an assumption: that physical influence resides in objects or states.

Faraday did not try to rescue that assumption. He did not introduce invisible intermediaries to explain the silence. He accepted the result at face value and allowed it to guide his thinking. Presence, he concluded, was not causal. At least, not in the way physics had assumed.

This realization forced a conceptual reset.

If static conditions produced no effect, then the cause had to lie elsewhere. It could not be located in the magnet as a thing, or in the current as a condition. It had to be located in what changed between conditions.

This insight did not yet explain induction—but it cleared the ground for it.

By recognizing the failure of presence, Faraday removed an entire class of explanations without argument. He did not disprove them philosophically; he let them expire experimentally. The apparatus simply refused to respond to stillness.

In UED terms—though Faraday would never phrase it this way—a system already in equilibrium has no reason to reorganize. Energy, evenly distributed and unperturbed, does nothing because there is nothing to resolve. Activity requires imbalance, and imbalance requires change.

Faraday did not yet speak of change as cause. But by 1831, he was certain of one thing:

Whatever induction was, it was not triggered by what *is*. It was triggered by what *ceases to be steady*.

That realization marked the true beginning of induction—not as a phenomenon, but as a new way of locating causality.

5.2 — Change as the Trigger

Once Faraday accepted that presence alone produced nothing, the experimental landscape reorganized itself almost automatically. If stillness was silent, then the only remaining candidate for causality was change. The task was no longer to intensify what existed, but to disturb it deliberately—and observe what followed.

The results were immediate and unambiguous.

When a current was switched on, a momentary current appeared in a nearby conductor. When the current was switched off, a current appeared again—opposite in direction. In between, while the current flowed steadily, nothing happened. The system responded not to the current itself, but to the transition into and out of it.

The same pattern appeared with magnets.

A stationary magnet, no matter how strong, induced nothing. But move the magnet toward a conductor, and a current appeared. Stop the motion, and the effect vanished. Reverse the motion, and the induced current reversed as well. Once again, the effect tracked change, not presence.

Rotation made the pattern even clearer.

When a conductor was rotated within the influence of a magnet, induction became continuous—not because something new was added, but because change was made continuous. Rotation was simply motion that never paused. It converted discrete change into sustained variation, and the system responded accordingly.

Across all these cases—switching, linear motion, rotation—the same rule held without exception:

- No change → no induction
- Change → response
- Continuous change → sustained response

The diversity of actions masked a deeper unity. Switching is temporal change. Motion is spatial change. Rotation is directional change. Different forms, same consequence. Induction did not care *how* change occurred. It responded only to the fact that conditions were no longer the same as they were a moment ago.

This was the breakthrough.

Faraday recognized that he was not observing multiple phenomena, but one phenomenon appearing under different disguises. The conductor was not reacting to electricity or magnetism as entities. It was reacting to variation in its environment. The system was sensitive not to states, but to differences between states.

This realization dissolved the need for separate explanations. There was no longer a "switching effect," a "motion effect," and a "rotation effect." There was induction—and induction was triggered whenever equilibrium was disrupted by change.

From a UED perspective, this is the moment where causality relocates decisively. Energy does not reorganize because something exists. It reorganizes because the existing distribution is no longer balanced. Change creates imbalance. Imbalance demands response.

Faraday did not yet formalize this as a law. He did something more decisive: he treated it as reliable behavior. He trusted it enough to stop asking whether induction would occur and began asking what kind of change produced what kind of response.

Induction, at this point, was no longer mysterious.

It was disciplined. It was conditional. And it obeyed a single, unforgiving rule:

If nothing changes, nothing happens.

5.3 — Why Motion, Switching, and Rotation Are the Same

At first glance, Faraday's induction experiments appeared to involve very different actions. Switching a current on and off felt electrical. Sliding a magnet felt mechanical. Rotating a conductor felt geometric. Each belonged to a different category of explanation, and early intuition encouraged treating them as separate causes with separate mechanisms.

Faraday refused that fragmentation.

He noticed that despite their surface differences, these actions produced identical consequences. A transient current appeared when a circuit was switched. A current appeared when a magnet moved. A sustained current appeared when rotation was maintained. The details varied; the response did not. The system behaved as if it were responding to a single underlying condition.

That condition was change itself.

Switching is change in time. Motion is change in position. Rotation is continuous change in orientation. These are not fundamentally different actions; they are different *expressions of variation*. Each replaces a static situation with a dynamic one. Each disrupts equilibrium in a way that cannot be ignored.

Seen this way, the unity becomes obvious.

Induction does not ask *what* is moving or *what* is being switched. It asks only whether the system is experiencing a difference from one moment to the next. If there is no difference, there is no response. If difference is introduced briefly, the response is brief. If difference is sustained, the response is sustained.

This insight dissolves a long-standing confusion. Motion is not special. Switching is not special. Rotation is not special. What is special is that all three refuse stillness. They deny the system the option of remaining the same.

Faraday's genius here was not in inventing a new category, but in removing unnecessary ones. He saw that physics had been over-partitioning behavior—assigning different causes to what was, in truth, a single pattern. By focusing on what the situations shared rather than how they differed, he simplified causality instead of multiplying it.

This is where the language of force finally collapses. Forces demand specificity: a direction, a magnitude, a point of application. Change demands none of these. It is indifferent to form. It operates wherever equilibrium is disturbed, regardless of the method used to disturb it.

For Faraday, this was the decisive clarification. He no longer needed to ask which action caused induction. He understood that induction was the system's generic response to change.

In UED-aligned terms, this is the recognition that systems respond to *delta*, not to objects. Redistribution is triggered by difference, not by identity. The pathway of difference may be temporal, spatial, or rotational—but the demand it places on the system is the same.

By unifying motion, switching, and rotation under a single principle, Faraday did more than explain induction. He redefined causality in electromagnetism. Effects were no longer anchored to things acting. They were anchored to situations changing.

From this point onward, electromagnetism could not be understood as a collection of forces. It had to be understood as a theory of responses to variation.

Change had become the cause.

5.4 — No Force, No Contact, No Substance

With induction firmly established as a response to change, the familiar language of force finally reached its breaking point. In Faraday's experiments, nothing was pushing anything in the ordinary sense. There was no contact, no collision, no identifiable agent transmitting effort from one object to another. And yet, a current appeared—reliably, predictably, and reversibly.

This absence was not a gap to be filled. It was the message.

Force-based explanations depend on contact or at least on a traceable intermediary. Something acts *here* and produces an effect *there*. Induction offered neither. The induced current did not originate in the magnet. It did not flow out of the source circuit. It appeared in a separate conductor, simply because conditions around it had changed.

No material substance traveled between the systems. No mechanical impulse crossed the gap. No push–pull story could be told without inventing invisible actors.

Attempts to rescue force language only made matters worse. To speak of "lines pulling charges" or "fields acting on conductors" risked mistaking description for mechanism. These phrases named regularities without explaining them. They gave comfort without clarity.

Faraday sensed this danger intuitively. He avoided treating induction as the result of something being transferred. He did not imagine electricity leaking from one circuit into another, nor magnetism exerting reach across space. He focused instead on what the system did when its surrounding conditions were altered.

This is the decisive shift.

Induction did not require contact because it was not about transmission. It was about reorganization. The system responded internally to a change in its environment. The cause was not an external agent acting upon it, but a loss of balance that demanded resolution.

In UED-aligned terms, this marks a clean separation between *interaction* and *redistribution*. No force needs to act if energy can reorganize itself locally in response to changing constraints. Induction is not something done to a conductor; it is something the conductor does when equilibrium is disturbed.

For Faraday, this realization required intellectual restraint. It would have been easy to invent a substance to carry influence, or a force to mediate action at a distance. Instead, he allowed the phenomenon to remain physically honest, even if the language lagged behind.

By refusing force, contact, and substance as explanatory crutches, Faraday preserved the integrity of the observation. He accepted that causality had shifted—from objects acting to systems responding.

This was not a small conceptual adjustment. It was a foundational one.

Once induction was understood without force, without contact, and without substance, electromagnetism could no longer be interpreted through mechanical analogy. It had become a theory of response to change, not a story of hidden pushes.

And with that, classical intuition finally gave way to something deeper—something that would soon demand a new language altogether.

5.5 — The System Responds, Not the Object

Induction made one fact impossible to ignore: the effect did not appear where the change originated. When a current was switched, or a magnet moved, the induced current emerged in a different conductor—one that had not been touched, pushed, or supplied with charge. The source changed; the response appeared elsewhere.

This alone dismantled object-centric causality.

If the cause resided in the object—magnet, wire, or source—then the effect should have been localized there. Instead, induction revealed a distributed response. The conductor that carried the induced current was not acted upon; it reorganized itself. The phenomenon belonged to the system as a whole, not to any single component.

Faraday noticed that the outcome depended decisively on configuration. Close a loop and induction appears; open it and nothing happens. Rotate the same conductor and the effect changes; alter orientation and the current reverses. The material remains identical, the environment similar—yet behavior shifts because the system's boundaries and geometry have changed.

This is the crucial reframe.

Induction does not ask what object is present. It asks how the system is arranged. The loop matters. The orientation matters. The boundary conditions matter. Cause is not located in a part; it is distributed across relationships.

From this perspective, the induced current is not something imposed on the conductor. It is the conductor's internal response to a change in its surrounding conditions. The system experiences imbalance and resolves it in the only way available—by establishing a current that counteracts the disruption.

In UED terms, this is energy redistribution operating at the system level. The system detects a delta—an imbalance introduced by change—and reorganizes locally to restore coherence. The response is not transferred; it is generated.

This insight explains why induction feels instantaneous yet non-contact. Nothing needs to travel from source to target. The system is already primed to respond. Change alters constraints; constraints alter internal distribution; current appears as the resolution.

For Faraday, this meant abandoning the search for an acting object altogether. He was not tracing influence from A to B. He was watching how B behaved when the overall situation changed.

Induction, then, is not an interaction between objects. It is a system-wide response to reconfiguration.

Once this is understood, the phenomenon stops being mysterious. The conductor does not need to be pushed. It needs to be situated within a changing arrangement. The rest follows naturally.

Causality had moved again—this time from change to systemic response.

5.6 — Reinterpreting Induction Under UED

When induction is viewed through the Universal Energy Dynamics (UED) framework, the phenomenon sheds much of its historical mystery. What appears, in classical language, as an induced current "caused" by motion or switching is more clearly understood as energy redistribution triggered by disrupted balance.

In UED terms, a system at rest is not inactive; it is balanced. Energy is already distributed in a way that satisfies its constraints.

Nothing changes because nothing needs to. The moment change is introduced—by motion, switching, or rotation—that balance is disturbed. The system is no longer in a state it can maintain.

Induction is the system's response to that disturbance.

The so-called "field" in this view is not an actor reaching out and doing work. It is a description of structured equilibrium—a way of representing how energy is arranged in space when nothing is changing. As long as that structure remains steady, it produces no observable consequence. It is only when the structure is altered that the system is forced to reorganize.

Change, therefore, is not an external trigger acting on the conductor. It is a reconfiguration of constraints. The conductor finds itself in a situation where its existing energy distribution is no longer compatible with the surrounding arrangement. A current appears not because something pushes charges, but because the system seeks a new balance.

This reframes induction completely.

The induced current is not evidence of energy arriving from elsewhere. It is evidence of local redistribution. The system absorbs the disturbance and responds internally, establishing motion of charge as the most efficient pathway back toward coherence. When change stops, redistribution stops. When change continues, redistribution continues.

Under this lens, induction is neither mysterious nor exceptional. It is a specific case of a general rule: whenever equilibrium is disrupted, energy reorganizes to restore it.

What makes induction special is not the rule, but the visibility of the response. Electrical systems make redistribution observable through current. Other systems redistribute energy through heat, deformation, or motion. The principle is the same; only the expression differs.

Faraday did not use this language, but his experiments behaved exactly as this framework predicts. He never needed to imagine forces acting across space or substances flowing invisibly. He watched systems reorganize when change made reorganization unavoidable.

Seen this way, induction is not about electricity discovering magnetism, or magnetism acting on conductors. It is about systems responding intelligently to disrupted balance.

That insight places Faraday's work far closer to modern dynamic thinking than to the mechanical metaphors of his time. He did not uncover a special trick of electromagnetism. He revealed a general principle of how structured systems behave when change enters the picture.

Induction, under UED, is simply balance in motion.

5.7 — Why Faraday Needed No Equation

By the time induction revealed itself fully in 1831, something unusual had already happened: the phenomenon was conceptually complete before it was symbolically expressed. Faraday did not wait for equations to tell him what induction was. He already knew—because the behavior was consistent, repeatable, and unified across situations.

This is why Faraday needed no equation.

Equations excel at compressing patterns once those patterns are understood. They are powerful summaries. But Faraday's work shows that compression is not discovery. Discovery happens when disparate observations collapse into a single explanatory idea. In induction, that idea was simple and unforgiving: *systems respond to change, not to presence.*

Faraday had already established this experimentally.

He had shown that switching, motion, and rotation were not separate causes. He had shown that no force, contact, or substance was required. He had shown that configuration and variation governed response. By the time mathematics entered the story, there was nothing left to invent—only something to formalize.

Later equations did not create induction. They described it.

This distinction matters because it reverses a common historical narrative. Faraday is often portrayed as the intuitive precursor whose ideas were later "completed" by mathematics. In reality, the intuition was already complete in the only sense that matters physically. The mathematics arrived to encode what was already clear at the level of behavior.

Faraday's notebooks did not lack rigor. They lacked symbolism by choice and circumstance. His rigor lived in repetition, variation, and coherence across experiments. If a phenomenon behaved the same way under different forms of change, that was evidence stronger than algebraic elegance.

From a UED perspective, this sequence is exactly what one would expect. Understanding must precede representation. Energy redistribution responds to real constraints, not to symbolic ones. Equations can mirror that response, but they cannot replace the insight that identifies what is responding and why.

Faraday trusted this order instinctively. He did not ask how to write induction down. He asked how reliably it appeared—and under what conditions it disappeared. Once those boundaries were clear, formal expression became optional rather than essential.

This is why induction feels inevitable in hindsight. It was never fragile. It did not depend on a clever derivation. It depended on a way of seeing that refused to confuse notation with necessity.

Faraday's greatest achievement here was not a law. It was restraint.

He allowed understanding to mature without forcing it into premature form. And by doing so, he ensured that when equations finally arrived, they described something real—something already anchored in the behavior of nature itself.

5.8 — The New Causality

With induction fully established, electromagnetism crossed a boundary from which it could not return. The shift was not merely technical; it was causal. What counted as a cause in physics had quietly but decisively changed.

Before induction, causality was anchored in objects. A magnet caused an effect. A current acted. A force was applied. Reality was explained by identifying *what* acted on *what else*. Stillness was assumed to be the default, and activity required an agent to disturb it.

Induction dismantled that picture without confrontation.

Faraday's experiments showed that objects could remain present, powerful, and perfectly inert. Nothing happened until something changed. Causality no longer resided in existence, but in variation. It was not the magnet that mattered, but whether its relationship to the system was altering. It was not the current that mattered, but whether it was beginning, ending, or reconfiguring.

Cause had moved from *things* to *transitions*.

This elevation of time, variation, and configuration transformed the conceptual foundation of physics. Static reality could no longer be treated as fundamental. The primary driver of effects was not what a system was, but what it was becoming. Change was no longer a secondary modifier; it was the initiating condition.

Equally important was the redistribution of responsibility within the system. Effects no longer needed to be traced back to a single acting entity. The system itself responded. Boundaries, orientation, and geometry became active participants. Causality became distributed, not localized.

This is why induction feels so modern.

It anticipates a view of nature in which balance, stability, and equilibrium are dynamic achievements rather than static givens. Systems are not passive recipients of influence; they are active resolvers of imbalance. When conditions shift, response is not optional—it is inevitable.

In this light, induction was not merely a unifying bridge between electricity and magnetism. It was a redefinition of explanation. It replaced force-driven narratives with response-driven ones. It replaced object-centered thinking with system-centered understanding.

From a UED perspective, this moment marks the recognition that reality operates through equilibrium-seeking behavior under constraint. Energy does not wait to be pushed. It reorganizes whenever balance is disturbed. Induction made this principle visible in one of its clearest forms.

Faraday may not have framed his work in these terms, but the implication was unavoidable. Once change became the cause, physics could no longer pretend that stillness was fundamental.

The universe, it turned out, was not built on rest with occasional interruptions. It was built on continuous adjustment, with stillness as a temporary condition.

Induction revealed that truth quietly—and permanently.

Closing Transition — Beyond Induction

Faraday did not stop when induction revealed itself. For him, induction was not a destination but a clarification—a confirmation that his long-held intuition about change was correct. Motion had shown that electricity could sustain activity. Induction showed something deeper: activity could arise wherever balance was disturbed, even without contact, force, or material transfer.

This realization closed one chapter and opened another.

Induction unified electricity and magnetism behaviorally, not by merging theories, but by revealing a shared rule of response. Systems did not care whether the disturbance came from a wire, a magnet, or a switch. They cared only that something had changed. That behavioral unity demanded a new way of speaking about space, influence, and continuity.

Out of this necessity, descriptive tools began to take shape. Faraday's "lines of force" were not substances filling space; they were attempts to remain faithful to what experiments were already saying—that influence was distributed, structured, and sensitive to configuration. These descriptions were born not from abstraction, but from restraint: a refusal to explain more than observation justified.

Yet this is also where risk enters the story.

As later thinkers sought to formalize Faraday's insights, description would gradually harden into ontology. What began as visual aids would start to sound like actors. What was meant to preserve intuition would sometimes replace it. Formalism would return—powerful, necessary, but once again capable of drifting ahead of meaning.

Faraday himself never crossed that line. He remained anchored to behavior. But the ideas he set in motion would travel farther than his method could control.

This closing transition matters because it marks a hinge in the history of electromagnetism. Induction resolved one confusion and created another challenge: how to formalize dynamic response without reintroducing static causes.

That challenge defines what follows.

The next part of this book will trace how Faraday's insights were carried forward—how fields emerged as language, how equations gained dominance, and where interpretation quietly slipped away from intuition. It will ask not only what was gained, but what was subtly misplaced along the way.

Once change became the cause, physics could no longer treat electricity as an event. But it also had to learn how to speak about change without turning it back into a thing.

That tension—between clarity and comfort—now takes center stage, with Faraday standing firmly on the side of clarity, even as the world moves on.

Interlude II

When Change Finally Had a Language

By the end of this part, something irreversible has occurred. Electricity is no longer an event. Magnetism is no longer a property. Causality is no longer anchored in objects.

Through motion and induction, Faraday did not merely add new phenomena to physics; he altered what physics was allowed to consider a cause. Change displaced presence. Configuration displaced substance. Systems replaced isolated actors.

This is not a small shift. It is a structural one.

Yet this is precisely where misunderstanding begins.

Once a discipline experiences such a shift, it faces a dilemma: how to talk about what it now understands without undoing the insight itself. Faraday had learned to see structure where others still searched for things. But language, especially scientific language, has a strong bias toward nouns, entities, and agents. It prefers *what exists* over *what reorganizes*.

This part showed Faraday at his clearest—working directly with behavior, guided by observation, unburdened by premature explanation. Next part begins where clarity meets interpretation, and where intuition is most at risk of being misread.

The contributions explored next are often treated as Faraday's most "mystical," most debated, or most paradoxical ideas. In reality, they are misunderstood for a simpler reason: they were never meant to be ontological claims.

- Lines of force were ways of *seeing structure*, not substances filling space.
- The Faraday disk exposed limits in force-based reasoning, not flaws in experiment.
- The Faraday cage demonstrated how boundaries govern redistribution, not how shielding "blocks" something.

In each case, the mistake is the same. What Faraday introduced as visual intuition was later absorbed as literal description. Tools meant to preserve understanding hardened into explanations that replaced it.

This interlude exists to slow the reader down before that hardening occurs.

Faraday did not invent entities. He invented *ways of seeing*. His genius was not in naming new things, but in refusing to name what could only be understood structurally. Where language threatened to mislead, he leaned on geometry, pattern, and behavior.

Next part revisits these contributions not to correct Faraday, but to correct how Faraday has been received.

We will not ask, *What did Faraday mean by lines of force as things?* We will ask, *What did Faraday see that required a new way of drawing?*

We will not ask, *Why does the Faraday disk violate equations?* We will ask, *What assumption about motion and change makes it look paradoxical?*

We will not ask, *How does the Faraday cage block influence?* We will ask, *How do boundaries reorganize energy so that nothing needs to be blocked?*

This part is uncomfortable for a reason.

It dismantles explanations that feel settled. It exposes how easily intuition is sacrificed for verbal comfort. And it shows that some of Faraday's most misunderstood ideas are misunderstood precisely because they remain too honest to be easily packaged.

If this part was about discovering that change is the cause, next part is about discovering why describing change correctly is harder than discovering it.

We now step into the most delicate territory of Faraday's legacy—not where he was wrong, but where we stopped listening carefully enough.

The tools are ready. The misinterpretations await.

Part III — The Most Misunderstood Contributions

He did not inherit explanations. He earned attention—patiently, repeatedly, and without permission

Chapter 6 — Lines of Force

Visualization as intuition, not ontology. Seeing Structure, Not Substance

The Drawings That Changed Physics

When Faraday reached for paper, he was not trying to illustrate a theory. He was trying to see.

His notebooks are filled with sketches that look almost disarmingly simple: curved lines around magnets, arcs looping through space, patterns traced again and again with quiet consistency. There are no equations asserting authority, no symbols declaring completion. Just lines—patient, repeated, deliberate.

At first glance, they appear informal. Almost naïve. In reality, they represent one of the most radical shifts in how physics learned to think.

Faraday was not drawing objects. He was drawing relationships.

Where others spoke of forces acting at a distance, Faraday drew how influence appeared to *occupy space*. Where others reduced interaction to point effects, he traced continuity. His lines were not claims about what space was filled with; they were records of how systems behaved when disturbed.

This distinction is everything.

Faraday did not believe there were threads in the air pulling iron filings into shape. He observed that iron filings *arranged themselves* into repeatable patterns and asked a deeper question: *What structure is being revealed by this response?* The lines were his answer—not as entities, but as visual summaries of behavior.

These drawings were revolutionary precisely because they refused to turn intuition into assertion. Faraday was careful. He never insisted that lines of force were physically real in the way objects are real. He treated them as guides for thought, not components of nature. They were scaffolding for understanding, not bricks of reality.

Yet history would not be so careful.

As electromagnetism advanced and mathematics rose to dominance, Faraday's drawings were reinterpreted. What he had meant as visualization was gradually absorbed as ontology. Lines became fields. Fields became things. What began as a way to *avoid* speculative entities became the seed of new ones.

This chapter begins here because the misunderstanding starts here.

If we misread Faraday's drawings as statements about substance, we lose what made them powerful. We turn intuition into doctrine. We replace structure with actors. And we inherit a century of confusion that Faraday himself never suffered from.

Faraday did not draw what exists. He drew what responds.

Before equations spoke fluently about fields, these lines taught physics how to look at space—not as empty background, but as structured by behavior. That lesson remains intact only if we resist the urge to literalize what was meant to clarify.

This opening scene sets the tone for the chapter that follows: not to defend Faraday's lines as real things, but to recover them as what they always were—

a way of seeing structure without inventing substance.

6.1 — Why Faraday Needed Lines at All

Faraday did not begin with lines because he preferred drawings to equations. He began with lines because nothing else available could faithfully represent what he was observing.

At the bench, effects did not appear at isolated points. They appeared across space. A magnet did not act only where it touched iron; it organized responses around itself. A current did not influence only the wire carrying it; it altered how nearby conductors behaved. Influence was extended, directional, and structured—but not localized.

Faraday needed a way to hold this structure in mind.

Without mathematical formalism to encode spatial variation, he faced a practical problem: how to represent distribution without inventing substance. Words were inadequate. Saying "the magnet acts" collapsed space into an object. Saying "force radiates" implied an agent. Both distorted what experiments were actually showing.

Lines solved this problem elegantly.

They allowed Faraday to represent direction, continuity, and density without committing to material existence. A line could show where response tended to align. A bundle of lines could show where response intensified. Curvature could show how geometry constrained behavior. None of this required assuming that anything was *there* in a substantive sense.

The lines were a bridge—not between electricity and magnetism, but between observation and understanding.

Experiments demanded a geometric explanation. Iron filings did not scatter randomly; they arranged themselves into stable, repeatable patterns. Coils responded differently depending on orientation. Boundaries shaped outcomes decisively. These were not properties of objects; they were properties of arrangement. Lines offered a way to track arrangement without turning it into an entity.

This is why Faraday did not speak first of fields, fluids, or forces. He spoke of lines of force because lines are relational by nature. They connect without occupying. They guide without acting. They describe without asserting existence.

In this sense, lines were an act of restraint.

Rather than inventing a mechanism he could not justify, Faraday chose a representation that stayed honest to the data. The lines did not explain *why* influence was distributed. They showed *how* it appeared to be distributed. That was enough.

Understanding could grow from there.

Faraday needed lines not because he lacked mathematics, but because he refused premature ontology. He needed a way to think spatially without pretending to know what space was filled with. Lines gave him exactly that—a way to see structure while remaining agnostic about substance.

They were not speculation. They were discipline made visible.

6.2 — What Faraday Actually Meant by "Lines"

When Faraday spoke of *lines of force*, he was not naming a thing. He was naming a pattern of response.

This distinction has been lost so often that it must be stated plainly: Faraday did not believe that space was threaded with invisible filaments, wires, or elastic strings. He was careful—almost cautious—in how he used the word *line*. For him, a line was a trace, not a substance; a record of how matter behaved when placed in a particular situation.

The classic example makes this unmistakable.

Scatter iron filings around a magnet, and they do not reveal what exists in space. They reveal how matter arranges itself under certain conditions. The filings respond to the situation they are placed in. Their alignment is an effect, not a revelation of hidden threads.

Faraday drew lines to summarize that alignment—nothing more, nothing less.

The lines therefore represent directionality of response, not direction of action. They show how a test object would tend to orient or move if introduced. They are predictive in a behavioral sense, not descriptive in a material one. No filing, no line. No response, no structure.

This is why Faraday emphasized density and curvature rather than endpoints. Where responses were stronger or more constrained, lines were drawn closer together. Where geometry redirected behavior, lines curved. These were not claims about pressure, tension, or flow. They were maps of tendency.

Crucially, Faraday never treated lines as agents. They did not pull. They did not push. They did not carry anything. They had no independent existence outside the behavior they summarized. The moment one imagines lines *doing* something, the meaning is already lost.

This restraint is what separates Faraday's use of visualization from later misreadings.

Faraday used lines to avoid speculative explanations, not to introduce them. Where others spoke of "electric fluids" or "magnetic virtues," he chose a representation that said, in effect: *this is how things behave; I will not pretend to know more than that.* The lines were honest because they stopped where observation stopped.

In modern terms, Faraday was using lines as behavioral gradients, not physical objects. They encoded orientation, not ontology. They answered the question *"how would a system respond here?"* without asserting *"what exists here?"*

This is why Faraday's lines remain powerful when read correctly—and misleading when read literally. They were never meant to populate space with entities. They were meant to discipline intuition.

Or stated simply:

Faraday did not draw what space contains. He drew how systems respond when space is disturbed.

That is the meaning of lines of force as Faraday intended them—and the meaning that must be recovered before anything that follows can make sense.

6.3 — Seeing Structure Instead of Inventing Entities

What set Faraday apart was not merely that he drew lines, but what he refused to do with them. Where many thinkers felt compelled to invent entities to explain influence, Faraday stopped at structure. He did not ask, *What is acting?* He asked, *How is the situation arranged?*

This difference is decisive.

An entity-based explanation looks for actors. Something must push, pull, carry, or transmit. Influence becomes a story about agents doing work. But Faraday's experiments did not support that narrative. They showed no identifiable source acting across space, no substance traveling from cause to effect. What they showed, consistently, was patterned response shaped by arrangement.

Faraday chose to see relational structure where others imagined hidden actors.

Structure, in this sense, is not a thing. It is an organization of conditions. It exists only insofar as behavior reveals it. Change the arrangement, and the structure changes. Remove the test matter, and the structure leaves no trace. Nothing persists independently of response.

This is why Faraday avoided force-carrier language. Words like "push," "pull," or "action at a distance" smuggled agency into situations where none could be experimentally located. They explained too much too quickly. Lines, by contrast, explained just enough. They captured orientation, continuity, and constraint without claiming an underlying substance.

Influence, for Faraday, was spatially distributed but not object-driven. A magnet did not *send* something outward. A current did not *emit* a force. Instead, the surrounding space—together with boundaries and configuration—defined how a system would respond if disturbed. The lines were a way to visualize that distributed sensitivity.

Seen correctly, Faraday's lines are snapshots of configuration, not pictures of reality's contents. They show how the system is poised to reorganize, where imbalance would propagate, and how constraints shape possible responses. They are conditional, not absolute. Situational, not substantive.

This approach required unusual restraint.

Inventing entities would have made explanation easier. It would have allowed Faraday to say, "This thing does that." But it would also have disconnected explanation from experiment. By refusing to cross that line, Faraday preserved a tight bond between observation and interpretation.

The result is subtle but profound: Faraday gave physics a way to see structure without pretending to know what structure is made of.

That discipline is precisely what later generations struggled to maintain. Once lines were treated as things rather than as visualizations, physics gained language but lost some clarity. Actors replaced arrangements. Ontology displaced intuition.

Faraday himself never made that move.

He visualized structure so that understanding could remain grounded in behavior. He drew lines not to populate space, but to prevent imagination from doing so prematurely.

In doing this, he demonstrated a principle that remains difficult even now: sometimes the most honest explanation is one that shows relationships and refuses to name entities.

That is not weakness. It is intellectual precision of the highest order.

6.4 — How Ontology Accidentally Crept In

The misunderstanding of Faraday's lines did not begin with Faraday. It began after him—quietly, gradually, and with good intentions.

Faraday's visual language was powerful precisely because it preserved intuition without asserting substance. But as electromagnetism matured and mathematics took center stage, a subtle pressure emerged: the demand to reify what had been drawn. Lines that once served as guides to perception were asked to carry explanatory weight. And once explanation is demanded, ontology tends to follow.

This shift was not malicious. It was methodological.

As equations began to describe electromagnetic behavior with increasing success, Faraday's lines were absorbed into formalism as if they referred to something *that exists*. The language changed almost imperceptibly. Lines that once *represented* influence began to be spoken of as if they *were* influence. Fields, originally descriptive constructs, hardened into entities that "fill space" and "act on charges."

Visualization quietly became literalization.

The problem was not mathematics itself. The problem was the reversal of roles. Instead of equations compressing an already-understood intuition, intuition was retrofitted to justify equations. The visual scaffolding Faraday had used to avoid speculative entities was now interpreted as evidence for them.

Language played a decisive role in this drift.

Phrases like "the field exerts a force" or "lines of force pull charges" sound innocuous, even helpful. But they smuggle agency into what Faraday had carefully kept relational. What was once a map of response becomes an actor in the story. Description becomes cause. Intuition becomes ontology.

Once this happens, confusion becomes inevitable.

Debates shift from *how systems respond* to *what fields really are*. Questions arise about whether fields are physical, whether they store energy, whether they propagate independently. Entire metaphysical structures are built around what began as a drawing meant to prevent metaphysics.

This is the irony.

Faraday introduced lines to avoid inventing invisible substances. Later generations treated those same lines as justification for inventing them.

Faraday himself never crossed this boundary. He remained careful in his language and modest in his claims. But once his work entered a culture increasingly comfortable with abstraction, the restraint that had guided the original insight was no longer enforced.

Ontology crept in not through argument, but through convenience.

It is easier to say "the field does this" than to say "the system reorganizes under these constraints." It is easier to imagine an actor than to hold a structure in mind. Over time, the easier story wins—not because it is truer, but because it is cognitively cheaper.

This section matters because it identifies the origin of many later paradoxes. When visualization is mistaken for substance, physics begins arguing with itself. Problems appear where none existed. Explanations multiply without clarifying behavior.

Recovering Faraday's intent requires undoing this drift—not by rejecting fields or equations, but by returning them to their proper role.

Lines were never meant to populate space. They were meant to discipline thought.

When that discipline is lost, intuition turns into ontology—and clarity quietly slips away.

6.5 — Lines of Force Under UED

When Faraday's lines of force are reinterpreted through the Universal Energy Dynamics (UED) framework, their true role becomes strikingly clear. They are not objects, carriers, or agents. They are contours of equilibrium—visual guides that indicate how energy is distributed *when nothing is changing*, and how it would redistribute *if something were to change.*

In UED, a stable system is not empty or inactive. It is balanced. Energy is already arranged in a configuration that satisfies existing constraints. Faraday's lines, in this sense, map that arrangement. They show where balance is tight, where it is loose, and where redistribution would preferentially occur if equilibrium were disturbed.

This immediately reframes several common misinterpretations.

Line density does not represent strength in the sense of force. It represents sensitivity to imbalance. Where lines are dense, a small disturbance produces a large response—not because something is pulling harder, but because the local configuration is more constrained. The system has fewer ways to absorb change quietly.

Line curvature does not indicate bending by an external agent. It reflects boundary conditions and geometry. The system is shaped by constraints—material, spatial, and relational—and the lines trace how those constraints guide redistribution pathways. Curvature is not action; it is limitation made visible.

Most importantly, lines do not describe motion. They describe readiness for motion.

In UED terms, Faraday's lines show how balance would attempt to restore itself when disrupted. They are conditional maps. They do nothing on their own because equilibrium requires no action. Only when change enters—motion, switching, variation—does redistribution begin, and the system "moves along" the pathways the lines already suggested.

This is why Faraday's lines were so predictive without being causal.

They did not act. They anticipated response.

Seen this way, lines of force are best understood as energy redistribution pathways—not in the sense of channels carrying energy, but as preferred directions in configuration space along which reorganization is easiest. They show how a system *would resolve imbalance*, not how something *acts upon it*.

This interpretation aligns perfectly with Faraday's restraint. He never claimed that lines existed independently. He treated them as summaries of behavior under constraint. UED simply gives that intuition a clearer conceptual footing: energy seeks equilibrium, and the structure of that equilibrium determines how redistribution will unfold.

Under this lens, Faraday's drawings regain their original power.

They are not metaphysical claims. They are not physical substances. They are visualizations of balance and constraint.

And this is precisely why they remain useful—provided we read them as Faraday intended. When lines are understood as maps of structured equilibrium, they clarify behavior without inventing actors. They preserve intuition without collapsing into ontology.

In UED terms, Faraday did not draw forces. He drew how balance is arranged in space.

That is why his lines still speak—quietly, accurately, and without pretending to be more than they are.

6.6 — Why Lines Are Still Useful (If Read Correctly)

Despite a century of misinterpretation, Faraday's lines of force remain one of the most effective cognitive tools in physics—but only when they are treated as visualization, not explanation.

Their value lies precisely in what they *do not* claim.

Modern equations are powerful, but they are abstract. They compress behavior into symbols that excel at prediction while often erasing spatial intuition. One can calculate accurately while having no sense of how a system is arranged, constrained, or poised to respond. Faraday's lines compensate for this loss. They restore geometric awareness—an understanding of orientation, continuity, and boundary-driven behavior that equations alone rarely provide.

Used correctly, lines function as cognitive scaffolding.

They allow the mind to hold a complex spatial relationship without turning it into a story about actors. They answer questions like:

- Where is the system most sensitive to change?
- How does geometry shape possible responses?
- Which configurations are stable, and which are fragile?

None of these questions require ontological commitment. They require pattern recognition, and that is exactly what lines facilitate.

This is especially important in teaching and conceptual development. Students often encounter electromagnetism as a collection of formulas detached from physical sense. Lines—when framed honestly—can reconnect mathematics to behavior. They give intuition somewhere to stand without misleading it into imagining invisible substances doing work.

The key condition is restraint.

The moment lines are described as things that "exist," "flow," or "act," their usefulness collapses. They stop clarifying structure and start competing with explanation. But when they are presented as maps of response, they regain their original strength. They show how systems are organized *before* change occurs and why certain responses follow *once* change is introduced.

In this way, lines serve a role similar to contour maps in geography. A contour line does not exist in the landscape. It represents elevation relationships. It helps one anticipate effort, direction, and constraint without claiming that lines are carved into the ground. Faraday's lines operate in exactly this manner for energetic systems.

Under UED-aligned thinking, their relevance becomes even clearer. Energy redistribution follows pathways determined by equilibrium structure and boundary conditions. Lines make those pathways intelligible without pretending to mechanize them. They allow intuition to track where balance will fail first and how resolution is likely to proceed.

This is why abandoning lines entirely would be a mistake.

The problem was never the lines. The problem was forgetting what they were for.

When read correctly, Faraday's lines do not compete with equations. They prepare the ground for them. They keep understanding anchored in spatial reality while mathematics does its work elsewhere. Together, they form a complete loop: intuition guides formalism, and formalism refines intuition.

Faraday understood this balance instinctively.

Recovering the usefulness of lines today does not require new theory. It requires restoring the discipline he practiced naturally: use visualization to see structure, and stop before imagination turns structure into substance.

Handled with that care, lines of force remain exactly what Faraday intended them to be— not answers, but guides to understanding.

6.7 — The Cost of Misreading Lines as Substance

When Faraday's lines of force were mistaken for physical substance, the damage was not immediate—but it was cumulative. What began as a subtle shift in interpretation gradually reshaped how entire generations thought about electromagnetism. The cost was not loss of predictive power; it was loss of conceptual clarity.

The first casualty was explanation.

Once lines were treated as things that exist in space, questions quietly changed. Instead of asking how systems respond to configuration and change, physics began asking what fields *are*, how they *store* energy, how they *act* on charges. Description slipped into agency. Visualization hardened into ontology. Behavior was no longer primary; entities were.

This produced an illusion of depth without increasing understanding.

Debates multiplied about the "reality" of fields, whether they are physical or mathematical, whether they propagate independently, whether they exert forces or merely encode them. These debates feel profound, but they arise largely from a category error: confusing a map with what the map was meant to summarize.

The second cost was pedagogical.

Students were taught to imagine invisible structures doing work in space, while simultaneously being told that nothing material was there. The result was cognitive dissonance. Lines that were meant to simplify intuition instead complicated it. Learners oscillated between pictures they were told not to take literally and equations they were told to trust without understanding.

Faraday never placed students in this position. The confusion entered later.

The third cost was conceptual fragmentation.

Once lines were reified, inconsistencies became unavoidable. Situations arose where the imagined behavior of "fields" conflicted with observed behavior of systems. Paradoxes appeared—not because nature was contradictory, but because explanation had drifted away from structure toward actors. The Faraday disk is one such case, where force-based and field-as-entity interpretations struggle to remain coherent.

These paradoxes are not failures of experiment. They are failures of interpretation.

Most damaging of all, misreading lines as substance shifted attention away from boundaries, geometry, and redistribution—the very elements Faraday was trying to make visible. Instead of asking how constraints shape response, physics often defaulted to asking how an invisible thing transmits influence. The system disappeared behind its supposed agents.

In UED terms, this is a predictable outcome. When equilibrium structure is mistaken for an active entity, energy redistribution is misattributed. What should be understood as balance-seeking behavior under constraint becomes a story about forces acting through space. The explanation feels complete but rests on an unnecessary invention.

Faraday avoided this trap instinctively.

He never claimed that lines act. He never claimed they carry anything. He never treated them as independent of the behavior they summarized. The later confusion is not his legacy—it is the result of forgetting the discipline with which the original insight was framed.

The cost of misreading lines as substance is therefore not merely historical. It persists wherever intuition is replaced by comfortable language, wherever visualization is mistaken for mechanism, and wherever explanation outruns observation.

Recovering Faraday's clarity does not require rejecting modern physics. It requires remembering why the lines were drawn in the first place.

They were never meant to explain what exists. They were meant to show how systems are arranged to respond.

When that distinction is lost, physics gains entities—but loses understanding.

Closing Transition — From Lines to Limits

Faraday's lines of force did their job the moment they revealed structure without inventing substance. They allowed influence to be seen without being personified, and response to be mapped without being mechanized. But lines, by their very nature, are open. They extend, curve, and distribute. They show continuity—but not containment.

And this is where Faraday's thinking moved next.

If lines reveal how a system is arranged, they also raise a sharper question: where does that arrangement stop? If influence is distributed, what defines its reorganization? If response follows structure, what role do boundaries play in stabilizing or redirecting that response?

Faraday did not answer these questions by adding new entities. He answered them by paying attention to limits.

The contributions that follow—often labeled paradoxical or misunderstood—are all consequences of taking lines seriously *without* reifying them. The Faraday disk exposes what happens when motion and redistribution are misread through force-based intuition. The Faraday cage reveals how boundaries reorganize energy without blocking anything. In both cases, the difficulty lies not in the phenomena, but in the assumptions brought to them.

Lines show structure. Boundaries govern resolution.

This transition matters because it marks a shift from visualization to constraint. From seeing how systems are arranged to understanding how they stabilize. Lines alone cannot explain why certain regions remain quiet while others respond violently, or why interior spaces can remain unaffected despite intense external activity.

To answer those questions, Faraday turned from drawing space to testing enclosure.

This part now moves into this terrain—the most misunderstood part of Faraday's legacy. It is here that intuition is most often replaced by slogans, and where explanation most often drifts into contradiction. The task ahead is not to complicate Faraday's ideas, but to remove the layers of misinterpretation that accumulated afterward.

We begin with the lines—not as things, but as guides.

We proceed to the limits—not as barriers, but as organizers.

Only then do Faraday's remaining contributions reveal what they always were: not anomalies in theory, but inevitable consequences of seeing structure clearly.

The next chapter follows the lines to where they fail— and shows that the failure was never Faraday's.

Chapter 7 — The Faraday Disk Paradox

Why flux-cutting fails and why textbooks still struggle

The Experiment That Refused to Behave

A simple disk. A magnet. A steady rotation.

Nothing about the setup looks provocative. No switching, no coils, no changing fields—just a conducting disk spinning calmly in a magnetic field. And yet, a voltage appears. Not briefly. Not transiently. Steadily.

This is where discomfort begins.

According to the story most students are taught, induction requires changing magnetic flux. No change in flux, no emf. The rule feels clean, reliable, and comfortably mathematical. But the disk does not cooperate. It produces an emf even when every reasonable accounting insists that the magnetic flux through the circuit remains constant.

The apparatus works. Reproducibly. Quietly. Without drama. What fails is the explanation.

This is why the Faraday disk has carried the label *paradox* for nearly two centuries—not because it is mysterious, but because it refuses to fit into a shortcut. It exposes a gap between what is calculated and what is actually happening.

What makes the situation more unsettling is that the experiment is not exotic. There is no delicate alignment, no hidden trick. It is robust enough to function as a generator. If the theory struggles here, the problem is not edge-case behavior—it is foundational interpretation.

Faraday himself did not see a paradox. He saw a clue.

He did not ask, "Why is flux changing?" He asked, "What is failing to settle into equilibrium?"

That question makes all the difference.

The disk does not depend on a loop whose enclosed area changes. It does not rely on cutting lines in the textbook sense. Instead, it sustains a persistent asymmetry—a configuration that cannot resolve itself while motion continues. Rotation keeps the system away from equilibrium, and the system responds exactly as it should: by redistributing energy.

The paradox, then, is not in the disk. It is in the expectation.

This opening scene matters because it marks the precise point where Faraday's intuition and later formalism begin to diverge. Flux rules work brilliantly in many situations—but they are summaries, not causes. When the summary is mistaken for the mechanism, experiments like the Faraday disk stop looking like confirmations and start looking like problems.

This chapter begins here, not to "fix" the disk, but to rescue it from misinterpretation.

The disk is not an exception to induction. It is induction revealed without shortcuts.

And that is why it still unsettles textbooks today.

7.1 — What Textbooks Expect to Happen

Most textbook treatments approach the Faraday disk with a quiet confidence that later turns into discomfort. The confidence comes from a rule that feels both elegant and reliable: induced emf equals the rate of change of magnetic flux. This formulation has been drilled into generations of students, reinforced by countless examples involving coils, loops, and changing fields. It works often enough that it begins to feel universal.

So expectations are set before the disk is even examined.

The reasoning typically goes like this: induction requires a changing magnetic flux through a closed loop. In the Faraday disk, the magnetic field is steady. The area of the disk is steady. The orientation is steady. Therefore, the flux is steady. And if the flux is steady, there should be no induced emf.

From this viewpoint, the disk should be electrically silent.

When it isn't, confusion follows—not because the observation is unclear, but because the rule has been promoted beyond its domain. A summary relation has been mistaken for a causal principle. What was meant to compactly describe many situations is now being asked to explain all of them.

Textbooks often try to patch this gap by stretching intuition. Some speak of "flux cutting" by moving conductors. Others invoke hidden loops or clever surface choices. Still others jump directly to force-based explanations without reconciling them with the flux narrative they just emphasized. The result is a collection of explanations that work locally but do not cohere conceptually.

The expectation itself is the problem.

Flux, in its textbook role, is a bookkeeping construct. It is a convenient way to summarize how field distribution and geometry combine over a chosen surface. It is not something the system measures, nor something nature tracks. Yet the expectation treats flux as if it were a physical trigger—something that must change in order for induction to occur.

That assumption holds only when the geometry conveniently aligns with loop-based thinking.

The Faraday disk does not offer that convenience. There is no single, well-defined loop undergoing change. The system is open, continuous, and distributed. The very idea of "the loop" becomes ambiguous. When the bookkeeping breaks down, the expectation collapses with it.

This is why the disk feels paradoxical only within the textbook frame.

The disk violates no physical law. It violates an overgeneralized shortcut. The expectation that induction must always announce itself as changing flux is not a law of nature—it is a pedagogical compression. When that compression is mistaken for causality, any configuration that lies outside its comfort zone looks like a problem.

In this sense, the Faraday disk is not difficult. It is honest.

It simply exposes the limits of an expectation that was never meant to be absolute—and invites a return to a deeper question: not *what changed in the equation*, but *what in the system failed to reach equilibrium*.

7.2 — Why Flux-Cutting Is a Derived Description

The idea of *flux-cutting* feels intuitive because it is easy to picture. A conductor moves, lines are crossed, and something happens. The language is vivid, mechanical, and reassuring. But this comfort hides a critical fact: flux-cutting is not a cause. It is a derived description—a convenient way to summarize certain outcomes after the physics has already occurred.

Flux itself is not something nature manipulates. It is a mathematical aggregate defined over a chosen surface. Change the surface, and the flux changes—sometimes dramatically—without anything physical having altered in the system. This alone should signal caution. Anything whose value depends on an arbitrary human choice cannot be the underlying trigger of a physical process.

In many textbook situations, flux language works because geometry is cooperative. A loop changes area, a field changes strength, and the bookkeeping aligns neatly with observation. Flux variation correlates with induced emf, so the correlation is promoted to causation. The shortcut is effective—but it remains a shortcut.

The Faraday disk removes that convenience.

There is no single loop whose area is changing. The disk is continuous. Charges move radially while the conductor rotates azimuthally. Any attempt to define a loop becomes artificial, and different choices lead to different flux narratives. The physics does not care which surface is chosen, but the flux calculation does. That mismatch exposes flux-cutting for what it really is: a descriptive overlay, not a physical mechanism.

This is why "cutting lines of force" fails as an explanation here. Lines, as Faraday intended them, were never objects to be cut. They were visual summaries of structure. Treating them as things that can be intersected by motion is already a category error layered on top of another—treating flux as causal.

What actually matters is configuration and motion, not line-crossing.

In the disk, rotation continuously maintains a non-resolvable asymmetry. Charges are constrained by geometry and boundaries in a way that prevents equilibrium. The system responds by redistributing energy, and an emf appears. This response does not wait for flux to change; it arises because balance cannot be achieved while the configuration persists.

From a UED perspective, this distinction is decisive. Flux is an accounting tool that sometimes mirrors redistribution behavior. It works when redistribution happens to align with loop-based geometry. It fails when redistribution is distributed, open, and boundary-driven, as in the disk.

This is not a flaw in Faraday's insight. It is a reminder of its depth.

Faraday did not say that induction is caused by cutting lines or changing flux. He showed that induction responds to change in configuration. Flux language was introduced later to package that insight efficiently—not to replace it.

The Faraday disk simply refuses to be packaged.

And in doing so, it teaches a lesson that textbooks often avoid: when a rule stops working, the mistake is rarely in nature. It is usually in mistaking a summary for a source.

7.3 — The Geometry of the Disk

The Faraday disk exposes a weakness that is easy to miss until geometry is taken seriously. The problem is not motion itself—it is how motion is arranged.

In most induction examples, geometry is loop-friendly. A wire forms a closed path. A magnetic field threads an area. Change occurs in a way that can be cleanly associated with a single loop. The mathematics is comfortable because the system is topologically simple.

The disk is not.

A rotating conducting disk has no privileged loop. Charges do not circulate around a fixed path; they drift radially while the material rotates azimuthally. Any "loop" drawn for analysis is imposed from outside, not dictated by the system. Different observers can draw different loops, choose different surfaces, and obtain different flux stories—without altering the physical behavior at all.

That alone is decisive.

If the explanation depends on which imaginary loop we prefer, then the loop is not fundamental. The disk's behavior is distributed and open, not confined to a single circuit. Geometry is continuous, not segmented. The system does not cooperate with loop-based intuition because it does not possess a loop in the causal sense.

What *does* it possess?

It possesses a persistent geometric asymmetry.

Rotation continuously displaces material relative to the magnetic configuration. The center and rim of the disk are not equivalent. Radial paths and azimuthal motion intersect constraints differently at different locations. This asymmetry is not transient; it is sustained as long as rotation continues. The system is never allowed to settle into a symmetric, equilibrium configuration.

This is why the disk produces a steady emf.

Nothing dramatic changes from moment to moment. There is no switching event to point at. But the configuration itself is incompatible with equilibrium. Geometry enforces a continuous imbalance, and the system responds continuously.

Loop-based thinking collapses here because it tries to compress a spatially extended, constraint-driven process into a single closed path. The disk resists this compression. It insists that induction be understood as a system-level response, not a loop-level effect.

This is also why explanations that jump between flux language and force language feel unsatisfactory. Each tries to localize what is fundamentally distributed. Each tries to assign cause to a place when the cause lies in arrangement.

The geometry of the disk teaches a hard lesson: induction does not require a changing loop. It requires a configuration that cannot resolve balance while motion persists. Where geometry enforces asymmetry, response follows—quietly, steadily, and without asking permission from our preferred diagrams.

The disk is not pathological. It is geometrically honest.

And once geometry is allowed to lead, the so-called paradox begins to dissolve—not through algebra, but through understanding how space, motion, and constraint cooperate to prevent equilibrium.

7.4 — Where Force-Based Explanations Break

When flux language begins to fail, textbooks often retreat to a familiar refuge: force. Charges move in a magnetic field, experience a Lorentz force, drift radially, and voilà—an emf appears. On paper, this sounds decisive. In practice, it only relocates the confusion.

The force-based narrative struggles here because it mixes kinematics with causality.

Yes, one can calculate forces on charges in the rotating disk. Yes, one can show that moving charges in a magnetic field experience deflection. But none of this explains why a steady emf persists without contradiction. It merely restates the motion in another language. The explanation becomes circular: charges move because forces act, and forces act because charges move.

More importantly, force language fragments the system.

It focuses attention on individual charges as if they were independent actors, responding locally to fields. But the Faraday disk is not a sum of isolated charge trajectories. It is a collective, constrained system. The emf is not a local push—it is a system-level outcome enforced by geometry, rotation, and boundary conditions.

Force-based explanations also introduce frame dependence that muddies causality. In one frame, the disk rotates and charges move. In another, the magnet rotates, or the observer rotates. Forces change description depending on perspective, yet the emf remains invariant. When an explanation shifts with the observer but the phenomenon does not, explanation has missed the cause.

This is the telltale sign of a patch.

Forces describe how motion appears once imbalance exists. They do not explain why imbalance is sustained. In the disk, the key fact is not that charges are deflected, but that the system's configuration never allows redistribution to complete. Rotation continuously reintroduces asymmetry. Forces merely express that asymmetry locally.

This is why the Lorentz-force narrative feels simultaneously correct and insufficient.

It calculates outcomes without identifying origin. It explains motion without explaining persistence. It answers *what happens* to charges while remaining silent on *why the system cannot settle.*

From a UED perspective, this is exactly what one would expect. Forces are derived effects, not primary causes. They emerge when energy redistribution encounters constraint. In the disk, the constraint is geometric and continuous. Force descriptions capture the local consequences of that constraint, but they do not define it.

Even Faraday did not lean on force here. He recognized that the disk's behavior could not be reduced to pushes and pulls without losing coherence. The phenomenon was not about acting forces; it was about maintained non-equilibrium.

The breakdown of force-based explanations in the Faraday disk is therefore instructive. It shows that whenever explanation requires switching frames, invoking multiple narratives, or patching with after-the-fact forces, the cause lies elsewhere.

Not in particles. Not in forces. But in configuration that refuses to relax.

Until that is acknowledged, force-based explanations will continue to describe the disk accurately—while explaining it poorly.

7.5 — What Is Actually Changing

The Faraday disk forces a deceptively simple realization: nothing that textbooks usually track is changing—yet something fundamental is.

The magnetic field is steady. The material properties are steady. The geometry of the disk is steady.

And still, the system refuses to settle.

What is changing is not a quantity, but a relationship.

In the rotating disk, the conductor is continuously re-entering a configuration that cannot equilibrate. Rotation does not create a sequence of events; it maintains a persistent mismatch between how energy would like to redistribute and how the geometry allows it to do so. The system is trapped in a state of *permanent adjustment.*

This is the crucial shift in perspective.

Induction here does not respond to time-variation in a field or flux. It responds to a non-resolvable imbalance—a configuration in which equilibrium is continuously postponed by motion. The system keeps trying to settle, and rotation keeps denying it closure.

In other words, what is changing is the availability of resolution.

At every instant, the disk presents the same structure, but the charges occupying that structure are never allowed to complete redistribution. The center and rim are not equivalent. Boundaries enforce asymmetry. Motion ensures that redistribution pathways are perpetually re-opened before balance is achieved.

This is why the emf is steady.

Not because something is repeatedly induced, but because induction never finishes.

From a UED standpoint, this is exactly how sustained response arises. Energy redistribution is the default tendency. When constraints prevent completion, the response persists. No oscillation is required. No switching is required. Only a configuration that cannot close its own imbalance.

This also explains why attempts to localize "the change" fail. There is no single moment, no privileged loop, no discrete event. The change is structural, not episodic. It exists at the level of the system, not at the level of variables.

Seen this way, the Faraday disk is not an anomaly at all. It is the cleanest possible demonstration that induction is not about changing fields, but about change in resolvability.

This is precisely the intuition Faraday was moving toward. He did not ask what fluctuated. He asked why balance never arrived.

The answer is now unavoidable:

Induction responds whenever a system is held away from equilibrium by configuration and motion— even if nothing "changes" in the way we are accustomed to measuring.

Once this is understood, the paradox evaporates. The disk does not violate Faraday's insight. It fulfills it—by revealing what induction was always responding to, but which our shortcuts taught us to overlook.

7.6 — Reinterpreting the Disk Under UED

Under the Universal Energy Dynamics (UED) framework, the Faraday disk stops being paradoxical the moment we stop asking the wrong question. The question is not *where is the changing flux?* or *which force acts on which charge?* The question is simpler and deeper: why is equilibrium never allowed to complete?

In UED terms, the rotating disk is a redistribution engine.

Energy, by nature, seeks balance. In a static configuration, redistribution proceeds until constraints are satisfied and the system settles. The Faraday disk is engineered—intentionally or not—to block completion of that process. Rotation continuously refreshes asymmetry. Geometry and boundaries prevent closure. The result is not oscillation, but persistent imbalance.

This reframing dissolves the paradox cleanly.

The emf is not caused by flux change; flux is merely a bookkeeping artifact that sometimes mirrors redistribution. The emf is the system's macroscopic response to a configuration that enforces ongoing redistribution without resolution. As long as the disk rotates, the imbalance remains non-resolvable, and the response remains steady.

Several consequences follow immediately:

- Motion is not the cause; it is the maintainer of imbalance.
- Fields are not actors; they describe the structured

equilibrium the system keeps failing to reach.
- Forces are local expressions, emerging where redistribution encounters constraint. They do not originate the effect.

From this view, the disk's behavior is inevitable. Rotation supplies continuous re-entry into an unresolved configuration. Boundaries (center vs rim, radial paths vs azimuthal motion) ensure that energy cannot redistribute symmetrically. The only stable system-level response is a sustained emf—exactly what is observed.

Notice what UED does *not* require.

It does not require special surfaces, hidden loops, or frame-dependent narratives. It does not require imagining lines being cut or forces being mysteriously summed. It requires only one principle: when equilibrium cannot be achieved, response persists.

This is why the Faraday disk is such a powerful confirmation of Faraday's original intuition. Long before equations hardened the story into flux-centric rules, Faraday recognized that induction followed change in *arrangement*, not change in *objects*. UED simply makes that recognition explicit and general.

Seen this way, the disk is not a special case. It is a clean case.

It demonstrates that induction is a global, configuration-driven phenomenon. It shows that steady outcomes can arise from steady conditions—provided those conditions prevent balance. And it confirms that the true driver is not a variable ticking in time, but a structure that denies closure.

Once induction is understood as equilibrium-seeking behavior under constraint, the Faraday disk stops asking for exceptions and starts offering insight.

7.7 — Why the Paradox Persists

The Faraday disk has been explained, re-explained, and "resolved" countless times—yet the sense of paradox refuses to disappear. This persistence is not due to experimental ambiguity or theoretical insufficiency. It survives because the wrong question keeps being asked, and the wrong explanatory habits keep being rewarded.

The first reason is pedagogical inertia.

Induction is almost universally taught through the lens of changing magnetic flux because it is compact, calculable, and easy to assess in exams. The rule works beautifully for coils and loops, so it becomes the rule. Once embedded, any configuration that does not fit is treated as an exception to be patched, rather than as evidence that the rule was never causal to begin with. The disk becomes a "special case," not a diagnostic.

The second reason is comfort with calculation over configuration.

Flux-based and force-based explanations both allow computation without requiring deep spatial reasoning. One can manipulate symbols and arrive at correct numbers while remaining conceptually agnostic—or confused—about what is actually happening. The Faraday disk demands something more difficult: holding geometry, motion, boundaries, and redistribution in mind simultaneously. That cognitive load is uncomfortable, so interpretation retreats to formulas that feel safer.

The third reason is the reification of explanatory tools.

Lines of force became fields; fields became actors. Flux became something that "changes" and "causes." Once these abstractions are treated as real agents, any phenomenon that does not align with their supposed behavior feels paradoxical. The disk exposes this mistake mercilessly. It shows that what was treated as a cause was only ever a summary of behavior under friendly geometry.

There is also a subtler reason: distributed causality is hard to teach.

The disk has no single locus of cause. There is no privileged loop, no moment of induction, no point where one can say, "here is where it happens." The cause is the *entire configuration*, maintained over time. Modern physics is comfortable with distributed effects in mathematics, but far less comfortable with them in narrative explanation. Textbooks prefer stories with actors and actions; the disk offers only structure and persistence.

Finally, the paradox persists because acknowledging it requires a quiet demotion of a beloved shortcut.

Accepting that "changing flux" is not the cause of induction feels like a loss—even though nothing predictive is lost. It feels as though clarity is being traded for ambiguity, when in fact the opposite is true. What is lost is a slogan; what is gained is coherence.

Faraday did not struggle with this paradox because he never elevated summaries into causes. He trusted configuration, boundary, and response. Later generations inherited his results but not always his restraint.

The Faraday disk continues to unsettle because it quietly insists on a different explanatory ethic: one that prioritizes how systems are arranged over which quantities are changing.

Until that ethic is restored, the disk will keep doing what it has always done— working flawlessly in the lab while refusing to be comfortable in the textbook.

The paradox survives not because it is unresolved, but because resolution demands letting go of an explanation that was never fundamental.

7.8 — What the Disk Teaches Us

Once the Faraday disk is stripped of its supposed paradox, it stops being an embarrassment to induction and becomes one of its cleanest teachers.

The first lesson is uncompromising: induction does not require changing magnetic flux. Flux change is a *frequent companion* of induction, not its cause. Where geometry happens to be loop-based and cooperative, flux bookkeeping mirrors reality well. Where geometry is continuous, open, and boundary-driven—as in the disk—flux loses its explanatory authority. Nature does not care which summary survives our notation.

The second lesson is equally sharp: loops are conveniences, not fundamentals. The disk teaches that insisting on closed-loop causality blinds us to distributed systems. Induction is not born at a point, along a wire, or around a loop. It is a system-wide response to a configuration that cannot resolve equilibrium. When there is no natural loop, forcing one into the picture only manufactures confusion.

Third, the disk teaches the priority of configuration over variables. Nothing "happens" in the dramatic sense. No field ramps up. No flux swings. No switch clicks. And yet the response persists. This tells us that induction is sensitive to *how elements are arranged and constrained*, not merely to how quantities evolve in time. Geometry, motion, and boundary conditions are not background details—they are causal.

Fourth, the disk reinforces Faraday's original intuition about change. Change is not always temporal variation of a measured quantity. Sometimes change is the denial of resolution—a configuration that is continuously refreshed so that balance is never achieved. Induction responds to that condition faithfully, without waiting for anything to "vary" in the textbook sense.

From a UED perspective, the lesson becomes even clearer. The disk demonstrates that sustained emf arises when energy redistribution is continuously demanded but never completed. Rotation is not a trigger; it is a maintainer. Geometry is not incidental; it is decisive. Boundaries do not block behavior; they organize it.

Most importantly, the disk teaches an ethical lesson about explanation.

Whenever a phenomenon works reliably but explanation feels strained, the problem is rarely with nature. It is usually with the story we insist on telling. The Faraday disk has been signaling this for nearly two centuries. It has never contradicted induction. It has only contradicted a shortcut.

Faraday did not see the disk as a special case or a loophole. He saw it as confirmation that induction follows configuration and change, not formulaic triggers. The discomfort entered later, when summaries hardened into causes and intuition was replaced by slogans.

Seen correctly, the Faraday disk does not undermine electromagnetism. It clarifies it.

It tells us that induction is broader, quieter, and more structural than we were taught to expect. It reminds us that when explanations fail, returning to arrangement, boundary, and response is not a retreat—it is a return to honesty.

And with that lesson absorbed, we are ready for the final step in this sequence: not motion, not redistribution, but containment.

The next chapter turns to enclosure—not as shielding, not as blocking, but as the ultimate test of whether we truly understand how systems reorganize themselves when boundaries are introduced.

The disk has done its work. It has taught us how not to misunderstand Faraday.

Closing Transition — From Redistribution to Enclosure

The Faraday disk does not end the story of induction. It tightens it.

By refusing to cooperate with flux shortcuts and force-centric narratives, the disk leaves us with a clarified principle: induction is a system-wide response to non-resolvable imbalance. Motion sustains redistribution. Geometry enforces asymmetry. Boundaries decide whether balance can ever be reached.

And that last word—boundaries—is where the discussion must go next.

The disk teaches us what happens when redistribution is continuously demanded and never allowed to finish. But it also raises a deeper, unavoidable question: *what happens when redistribution is allowed to finish everywhere except one place?* What happens when geometry does not merely frustrate equilibrium, but protects an interior from participating in it?

This is not a minor extension. It is the logical continuation.

If lines of force show structure without substance, and the disk shows redistribution without flux change, then enclosure becomes the ultimate test. Can a system reorganize itself so completely that internal regions remain unaffected—even while external activity is intense? Can stability emerge not by blocking influence, but by redistributing it around a boundary?

This is where misunderstanding peaks.

The phenomenon most often invoked to answer these questions—the Faraday cage—is routinely described in the wrong language. It is said to "block fields," "shield charges," or "cancel forces." Each phrase repeats the same mistake the disk exposed: treating descriptive tools as acting agents.

Faraday did not think this way.

For him, enclosure was not about stopping something from entering. It was about allowing the system to resolve imbalance at the boundary so completely that the interior had nothing left to do. Stability was not enforced; it was achieved. Silence inside the cage was not imposed—it was earned.

The transition from disk to cage is therefore not a change of topic. It is a change of resolution scale. The disk showed how motion sustains imbalance. The cage shows how boundaries absorb it. One reveals why response persists; the other reveals how response can disappear without contradiction.

Together, they dismantle the last refuge of force-based intuition.

The next chapter does not ask how influence is blocked. It asks how redistribution completes itself so perfectly that nothing remains to propagate inward.

If the disk exposed the limits of flux, the cage will expose the limits of shielding metaphors.

And in doing so, it will bring Faraday's intuition to its quiet conclusion: that stability, like induction, is not a matter of force—but of arrangement, boundary, and completion.

We now move from motion to enclosure— from persistent imbalance to earned stillness.

Chapter 8 — The Faraday Cage

Boundaries, redistribution, and internal stability

Silence Inside a Storm

The moment is deceptively simple.

A metal enclosure stands exposed while electrical activity rages around it—sparks crackle, charges accumulate, tension builds on the outside. Every intuition trained on force and action expects the interior to react. Something must leak in. Something must get through.

Nothing does.

Inside the enclosure, instruments remain calm. No deflection. No spark. No disturbance. The interior is quiet, not delicately balanced, not temporarily spared—but fundamentally unaffected.

This silence is what unsettled early observers.

If electricity acts at a distance, why does distance suddenly fail? If forces propagate through space, why do they stop at a thin boundary? If fields "fill space," how can space inside remain untouched?

The usual answers arrive quickly—and misleadingly. The cage is said to *block* fields. Charges are said to *cancel* influence. The enclosure is described as a defensive barrier, a wall against nature's reach. The language feels intuitive, almost comforting.

But the experiment itself offers no evidence of a battle.

There is no sign of struggle at the boundary. No buildup inside that must be neutralized. No delayed response waiting to be suppressed. The interior is not protected—it is irrelevant to what is happening outside.

This is the first clue.

Faraday did not interpret this calm as the success of shielding. He treated it as evidence of completion. Something about the enclosure allowed all redistribution to resolve itself *before* anything could propagate inward. The boundary was not resisting influence; it was absorbing the need for it.

The storm outside was real. The silence inside was earned.

This opening scene matters because it reveals the same pattern we have seen before—misread through habit. Just as lines were mistaken for substances, and the disk was mistaken for an exception, the cage has been mistaken for a blocker. In each case, the metaphor survives because it sounds active. In each case, Faraday's insight was quieter.

The Faraday cage does not stop electricity. It finishes its work.

This chapter begins here, at the moment where intuition expects action and instead finds resolution. To understand why the interior is calm, we must stop asking what is being blocked and start asking where imbalance is allowed to complete itself.

Only then does the silence make sense—not as absence, but as achievement.

8.1 — The Popular Story: Shielding as Blocking

The most common explanation of the Faraday cage is also the most misleading. It is said to *block* electric fields, *stop* influence, or *shield* whatever lies inside from what happens outside. The language is vivid and reassuring. It borrows from everyday experience—walls stop wind, armor stops blows, insulation stops heat. By analogy, the cage must stop electricity.

This story persists because it feels active.

Blocking suggests resistance. Shielding suggests defense. Cancellation suggests opposing forces battling to a draw. Each metaphor assigns agency to the enclosure and drama to the process. Something tries to get in; something else prevents it. The interior is safe because the boundary fights on its behalf.

But this narrative collapses under even light scrutiny.

First, nothing in the experiment behaves like a blocker. There is no sign of accumulation pressing inward. No delayed seepage. No internal reaction that must be suppressed. The interior is quiet from the outset, not because something is continually held back, but because nothing ever arrives that needs to be stopped.

Second, "blocking" presumes an incoming influence that travels toward the interior. Yet Faraday's observations showed no such propagation attempting to cross the boundary. What happens instead is rearrangement at the surface. Charges move, redistribute, and settle—entirely at the enclosure. The process completes there. The interior never becomes part of the story.

Third, the blocking metaphor quietly reintroduces the very ontology Faraday avoided. It treats fields as projectiles, forces as agents, and space as a battleground. The cage becomes a wall against invisible things. This framing does not arise from observation; it arises from habitual language.

Even the word *shield* is revealing. It implies protection against harm. But nothing harmful is detected inside, not even briefly. The enclosure does not protect the interior from instability; it prevents instability from becoming relevant to the interior in the first place.

This is why the popular story creates more confusion than clarity. It answers a question Faraday was not asking—*how is influence stopped?*—instead of the one his experiments demanded—*where does imbalance resolve?*

Faraday never described the cage as a blocker. He described what he saw: charges rearranged on the outside, and the inside remained unaffected. The explanation followed observation, not metaphor. The calm inside was not enforced; it was achieved through redistribution.

The popularity of the blocking story is therefore understandable but costly. It replaces completion with opposition, structure with struggle, and resolution with resistance. It feels intuitive—but intuition here is being misled by language.

To move forward, the metaphor must be dropped.

The Faraday cage does not work because it blocks something. It works because it allows everything that needs to happen to finish at the boundary.

Once that is understood, the silence inside stops being mysterious—and starts being inevitable.

8.2 — What Faraday Actually Observed

Faraday did not begin with an explanation. He began with patience.

When he placed instruments inside a conducting enclosure and subjected the exterior to strong electrical activity, he did not look for subtle residual effects or delayed responses. He looked for *any* sign that the interior participated at all. Again and again, it did not.

No deflection appeared on electroscopes. No spark jumped. No gradual buildup hinted at a struggle being contained.

The interior was not partially affected. It was categorically uninvolved.

This observation mattered more to Faraday than any metaphor. If the cage were "blocking" influence, one would expect evidence of pressure at the boundary or traces of suppressed activity inside. But Faraday saw nothing of the sort. The behavior was clean: charges moved and settled on the outer surface, and once that redistribution completed, the experiment was over.

Crucially, Faraday did not infer invisible counter-forces or hidden cancellations. He did not imagine the enclosure fighting external influence. He described what actually happened: redistribution occurred where it was allowed to occur, and nowhere else.

The surface was active. The interior was irrelevant.

This distinction is subtle but decisive. Faraday recognized that the enclosure did not act as a barrier to something trying to pass through. Instead, it acted as a site of completion. All imbalance introduced from outside found its resolution at the boundary. Once resolved, there was no remaining reason for the interior to respond.

Faraday also noted that this behavior was robust. It did not depend on delicacy, precise tuning, or special materials beyond conductivity. The phenomenon was not fragile. That robustness itself was evidence against force-cancellation stories, which tend to be sensitive to alignment and magnitude. What Faraday observed was structural, not contingent.

Even more telling was what Faraday did *not* observe.

He did not observe accumulation inside waiting to be neutralized. He did not observe delayed equilibration. He did not observe any dependence on distance from the boundary.

The interior behaved as if it were not part of the electrical system at all—because, functionally, it wasn't.

This is why Faraday treated the cage as an insight into how systems close themselves, not as a trick for shielding. The enclosure simply defined where redistribution was permitted to finish. Once finished, the interior had no imbalance to address.

Faraday trusted this observation enough to let it stand without embellishment. He resisted the temptation to dramatize it with agents, barriers, or counteracting forces. The calm inside the cage was not defended. It was earned by completion elsewhere.

Understanding what Faraday actually observed is essential, because it strips away a century of inherited metaphor and returns us to the experimental core. The Faraday cage is not mysterious when described accurately. It is only mysterious when described as something it never was.

Faraday did not observe blocking. He observed resolution at the boundary and irrelevance of the interior.

Everything that follows depends on keeping that distinction intact.

8.3 — Boundaries as Active Organizers

The decisive mistake in most explanations of the Faraday cage is treating the boundary as passive—a wall that merely stands in the way. Faraday's experiments reveal the opposite. The boundary is not inert. It is where the system organizes itself.

A conducting enclosure does not wait for influence to arrive and then decide whether to let it pass. The moment external imbalance is introduced, the boundary participates. Charges move, redistribute, and settle along the surface in a way that satisfies the constraints imposed by geometry and conductivity. This process is not defensive; it is resolutive.

Boundaries, in this sense, are not barriers. They are decision points.

They define where energy redistribution is allowed to occur and where it is not required. The enclosure's surface becomes the natural site where imbalance can fully express itself and complete its adjustment. Once that completion happens, there is nothing left to propagate inward.

This is why the surface is active while the interior is silent.

The geometry of the boundary matters profoundly. Shape, continuity, and connectivity determine how redistribution unfolds. A smooth, closed surface allows imbalance to spread and resolve without leakage. Discontinuities or openings change the story—not because something "gets through," but because completion is no longer possible at the boundary alone.

Seen this way, the boundary organizes response rather than resisting it.

This organizing role explains several features that blocking metaphors cannot:

- Why charges accumulate precisely where they do
- Why the interior shows no transient disturbance
- Why the effect is robust across magnitudes and configurations
- Why no special thickness or strength is required beyond conductivity

The boundary does not oppose influence. It absorbs the need for further influence by completing redistribution at the earliest possible stage.

From a UED perspective, this behavior is expected. Energy seeks equilibrium. When constraints allow equilibrium to be achieved at the boundary, the system stops there. The interior is not protected; it is simply no longer relevant to the process.

This is the same principle seen earlier in a different form. In the Faraday disk, geometry prevents completion, so response persists. In the Faraday cage, geometry enables completion, so response terminates.

The difference is not force. It is where the system is allowed to finish its work.

Faraday understood this intuitively. He did not speak of boundaries as shields, but as conditions. The enclosure did not fight electricity; it gave electricity a place to settle completely.

Once that is understood, the Faraday cage stops looking like a wall and starts looking like what it truly is: a structural organizer of equilibrium.

And with that shift, internal stability is no longer mysterious—it is simply the natural consequence of letting redistribution finish where it belongs.

8.4 — Redistribution, Not Cancellation

One of the most persistent myths surrounding the Faraday cage is the idea of cancellation—that opposing fields or forces somehow neutralize each other, leaving the interior unaffected. This language sounds technical, even precise, but it quietly misrepresents what actually occurs.

Cancellation implies conflict.

It suggests that something enters the enclosure, meets an equal and opposite influence, and is nullified. Two actors collide, and the result is zero. This framing assumes incoming influence, internal opposition, and a balancing act taking place *inside* the system.

None of this is observed.

Faraday did not see opposing effects battling to a draw. He saw redistribution completing itself at the surface. Charges moved until no further adjustment was required. When equilibrium was reached at the boundary, the process ended—not because something was canceled, but because nothing remained unresolved.

Cancellation language is tempting because it feels mathematically neat. Zero field inside. Net effect equals zero. But this is a result, not a mechanism. Saying "it cancels" merely restates the outcome in reverse. It does not explain how or where the system reorganized to make that outcome inevitable.

Redistribution, by contrast, explains both.

Energy introduced externally seeks equilibrium. In a conducting enclosure, the easiest place for that equilibrium to be achieved is the surface. Charges are free to move there. Geometry allows full reorganization. Once redistribution completes, the internal region has no imbalance to respond to. Silence follows naturally.

No internal forces are required. No opposing agents are invoked. No invisible tug-of-war takes place.

This distinction matters because cancellation language subtly reintroduces force-based thinking. It encourages the belief that something must be actively suppressed. Redistribution-based thinking shows that suppression is unnecessary when resolution is complete.

From a UED perspective, this is a textbook case of equilibrium-seeking behavior. Systems do not negate influences; they reorganize themselves until imbalance disappears. The Faraday cage is effective not because it negates electricity, but because it provides a complete and sufficient pathway for resolution at the boundary.

This also explains why the interior shows no transient effects. If cancellation were occurring internally, one might expect momentary disturbances before balance is restored. Faraday observed none. The interior was quiet from the start, because it was never part of the redistribution process.

Faraday did not frame his observations in terms of cancellation for this very reason. He described movement, settlement, and completion—never opposition. His restraint preserved clarity where later metaphors introduced confusion.

The Faraday cage does not cancel influence. It finishes redistribution so thoroughly that no influence remains to propagate.

Once this is understood, internal calm stops looking like a mathematical trick and starts looking like a structural consequence—exactly as Faraday observed it.

8.5 — Internal Stability as an Earned State

The calm inside a Faraday cage is often described as protection, as if the interior were fragile and in need of defense. This framing is backwards. The interior does not *need* protection because, structurally, it never enters the problem.

Internal stability is not imposed. It is earned.

Once redistribution completes at the boundary, the interior is left with nothing to resolve. There is no residual imbalance, no gradient, no unfinished adjustment demanding participation. In energetic terms, the interior is already at equilibrium—not because it was shielded from change, but because change was fully accounted for elsewhere.

This is a crucial distinction.

If the interior were being protected by active suppression, we would expect evidence of effort—temporary disturbances, corrective forces, delayed settling. None are observed. The interior remains quiet because the system has no reason to involve it. Stability is not maintained by vigilance; it emerges from completion.

This also explains why the material inside the cage does not matter. Objects placed within do not alter the outcome unless they disrupt the boundary's ability to complete redistribution. The interior is dynamically neutral. It neither attracts nor resists influence because there is no imbalance left to drive response.

In UED terms, the Faraday cage demonstrates a simple rule: when equilibrium is achieved at the lowest-resistance location available, the rest of the system becomes irrelevant to redistribution. The boundary does the work so completely that the interior never becomes part of the energetic pathway.

This reframes stillness itself.

Stillness is not absence of activity. It is absence of requirement. The interior is calm not because nothing could happen there, but because everything that needed to happen has already happened at the surface.

This insight generalizes beyond electromagnetism. In any system—physical, biological, or even social—stability inside a boundary arises when pressures are resolved at the edges. When boundaries absorb imbalance efficiently, interiors remain coherent without enforcement. Control gives way to structure.

The Faraday cage is therefore not a lesson in shielding. It is a lesson in where equilibrium is allowed to finish.

Internal stability is not defended. It is made unnecessary to disturb.

And that is why the silence inside the cage is so complete. It is not a victory over external influence; it is the natural outcome of a system that has already settled its accounts.

8.6 — Reinterpreting the Cage Under UED

Under the Universal Energy Dynamics (UED) framework, the Faraday cage is not a special device with a special power. It is a natural consequence of how equilibrium-seeking systems behave when boundaries are introduced correctly.

UED begins with a simple premise: energy redistribution continues until equilibrium is achieved, and it always follows the lowest-resistance path permitted by constraints. The Faraday cage is a textbook realization of this principle. It does not alter energy's intent; it merely defines where that intent can complete itself.

In a conducting enclosure, the boundary offers the most permissive redistribution pathway. Charges are free to move along the surface. Geometry allows imbalance to spread, reorganize, and settle without obstruction. Once this redistribution reaches completion at the boundary, the system has exhausted its need to act. There is no leftover gradient demanding further response.

The interior, therefore, is not protected—it is bypassed.

From a UED standpoint, the cage functions as a redistribution terminus. All externally introduced imbalance finds its resolution there. The interior becomes dynamically irrelevant not because it is isolated, but because it is no longer required for equilibrium completion. Energy does not propagate inward because propagation would serve no purpose.

This also clarifies why material thickness, strength, or "shielding power" are secondary. As long as the boundary remains conductive and continuous enough to allow full redistribution, the effect holds. Once that condition is satisfied, adding more material does not increase protection—because nothing is being resisted.

UED also explains why imperfections matter. Gaps, discontinuities, or insulating breaks in the enclosure do not "let influence leak through" in an active sense. They simply prevent redistribution from completing at the boundary. When completion fails, the system searches for other pathways—including the interior. What appears as leakage is actually unfinished work.

Seen this way, the Faraday cage is not an electromagnetic anomaly. It is a clean demonstration of a general rule: stability emerges wherever equilibrium is allowed to finish locally.

This interpretation is fully consistent with Faraday's experimental restraint. Faraday did not invoke special forces, invisible barriers, or compensating agents. He trusted that if redistribution was complete at the surface, the interior would remain silent—and it always did.

Under UED, the Faraday cage is therefore elevated from a clever trick to a structural principle. It shows that boundaries are not passive edges, but active sites of resolution. When designed to allow completion, they eliminate the need for deeper intervention.

The cage does not block energy. It gives energy nowhere further to go.

And in UED terms, that is the most stable condition a system can achieve.

8.7 — Why the Misinterpretation Persists

The Faraday cage is one of the most robust and reproducible demonstrations in physics—and yet it remains one of the most persistently misunderstood. This is not because the phenomenon is subtle, but because the wrong explanatory instincts keep reasserting themselves.

The first reason is language bias.

Human explanation is biased toward verbs with actors: block, stop, cancel, repel. These words imply effort, resistance, and confrontation. They feel explanatory because they mirror everyday experience. A wall stops a ball. An umbrella blocks rain. When faced with a quiet interior and a noisy exterior, the mind reaches for the same vocabulary. The enclosure must be *doing something*.

But Faraday's experiments show no struggle. The calm is effortless. Language, not observation, introduces drama.

The second reason is pedagogical convenience.

"Shielding" is easy to teach. It compresses a complex redistribution process into a single word that sounds decisive. Students can memorize it, repeat it, and move on. Explaining boundary-driven completion, by contrast, requires spatial reasoning, patience, and comfort with distributed causality. That is harder to package—and harder to test in an exam.

So the shortcut survives.

The third reason is ontological inertia.

Once fields are treated as things that exist and act, it becomes natural to imagine them being blocked or excluded. The cage is then forced into the role of a gatekeeper. This reification makes the phenomenon feel intuitive while quietly disconnecting it from what Faraday actually observed.

The fourth reason is confusion between result and mechanism.

"Yes, the field inside is zero." "Yes, the net effect is canceled."

Both statements are true—but neither explains *how the system arrived there.* They describe the endpoint and mistake it for the process. When the endpoint is re-described as a mechanism ("it cancels"), explanation stalls.

Finally, the misinterpretation persists because the correct explanation feels passive.

Redistribution completing at the boundary does not sound heroic. It does not assign agency. It does not dramatize causality. It simply says: the system settles where it can. For many, this feels unsatisfying—even though it is precisely what the experiment shows.

Faraday was comfortable with this quietness. He trusted outcomes that required no spectacle. Later generations inherited his results but often replaced his restraint with metaphors that felt more active, more intuitive, and ultimately more misleading.

The misinterpretation of the Faraday cage persists not because physics lacks answers, but because we keep preferring stories that sound like action over explanations that describe completion.

Until that preference is questioned, the cage will continue to be described as a shield— even though it has never behaved like one.

Understanding requires letting go of the urge to dramatize and accepting a subtler truth: that the most effective boundaries do not fight influence— they simply leave it nowhere else to go.

8.8 — What the Cage Teaches Us About Stability

The Faraday cage delivers one of the most quietly radical lessons in physics: stability is not imposed by resistance; it emerges from completion.

What the enclosure demonstrates is not how to stop influence, but how to finish redistribution so thoroughly that further response becomes unnecessary. Stability, in this sense, is not an active state maintained by force. It is a structural outcome—the natural stillness that appears once a system has exhausted its need to reorganize.

This reframes stability itself.

In the common imagination, stability is something defended. Walls are built. Shields are raised. Forces oppose forces. But the Faraday cage shows a different principle at work. The interior is stable not because it is guarded, but because all imbalance has already been resolved at the boundary. There is nothing left to disturb it.

This lesson generalizes far beyond electromagnetism.

Any system—physical, biological, social, or cognitive—achieves internal coherence when pressures are allowed to resolve at the edges. When boundaries absorb imbalance efficiently, interiors remain ordered without control. When boundaries fail, interiors are forced to participate, and instability follows. The Faraday cage is simply a clean, visible instance of this universal pattern.

It also teaches us where to look when stability fails.

If the interior becomes noisy, the problem is not that influence "leaked in." The problem is that completion did not occur at the boundary. Gaps, discontinuities, or constraints that prevent full redistribution shift the burden inward. Instability is not invasion; it is unfinished work.

From this perspective, the Faraday cage is not a special object with a special effect. It is a demonstration of how equilibrium-seeking systems behave when given the right conditions. Conductivity, continuity, and geometry are sufficient. No special forces are required. No exceptional materials are needed. The principle is structural, not exotic.

Faraday grasped this intuitively. He did not celebrate the cage as a triumph of shielding technology. He treated it as evidence that nature, when allowed to reorganize freely at the right place, produces stillness on its own.

The deeper teaching, then, is not about electricity at all.

It is about how calm arises.

Stability is not achieved by fighting disturbance. It is achieved by giving disturbance a place to finish.

The Faraday cage shows this with uncompromising clarity. It completes the arc that began with lines (structure), passed through the disk (redistribution), and ends here—with earned stillness.

And in doing so, it leaves us with a principle that outlives any single experiment: that the most resilient systems are not those that resist change most aggressively, but those that resolve it most completely, at the right boundary.

Closing Transition — Faraday's Quiet Conclusion

With the Faraday cage, a pattern that has been unfolding across this part of the book reaches its natural end.

First came structure, revealed through lines of force—not as substance, but as arrangement. Then came redistribution, exposed by the rotating disk—not as flux change, but as non-resolvable imbalance. Finally comes completion, demonstrated by the enclosure—not as blocking, but as equilibrium fully achieved at the boundary.

Nothing new was added at each step. Only misunderstanding was removed.

This is why Faraday's contributions feel deceptively simple when seen clearly. He did not invent actors to explain effects. He did not multiply causes when one structural principle was sufficient. He trusted that if observation was followed honestly, explanation would remain modest—and therefore correct.

The cage is Faraday's quiet conclusion.

It shows that stability is not enforced by opposition, not maintained by constant effort, and not achieved by suppressing influence. Stability appears when a system is allowed to finish its reorganization at the right place. Once that happens, silence is not fragile. It is absolute.

This insight closes Faraday's experimental arc.

Lines taught us how to see structure without substance. The disk taught us how motion sustains imbalance without flux change. The cage taught us how boundaries allow imbalance to disappear without force.

Together, they form a coherent worldview—one that never required mystical fields, invisible blockers, or dramatic cancellation. Only arrangement, constraint, and completion.

What follows in the book now moves beyond Faraday—not to correct him, but to examine what happened *after* him. How his restraint was gradually replaced by convenience. How visual intuition hardened into ontology. How equations gained authority while explanation quietly drifted.

But Faraday himself needs no rescue.

Faraday already left us with everything required: a way of thinking that favors structure over story, completion over confrontation, and understanding over comfort.

The work ahead is not to add to Faraday. It is to listen to him without translating his clarity into something louder than it needs to be.

With that, this part closes—not with a new principle, but with a restored one:

That nature does not fight imbalance. It simply resolves it—and when it is resolved fully, stillness follows.

From here, the book is ready to step outward—from Faraday's intuition to the broader consequences of how we choose to explain the world.

Interlude III

When Clarity Became Too Powerful

By the end of this part, something subtle but decisive has happened.

Nothing new was added to nature. Nothing exotic was discovered. And yet, understanding became dangerously complete.

Through Faraday, we saw a full arc unfold—without excess language, without invented actors, without metaphysical ambition. Structure was revealed without substance. Redistribution was explained without flux mysticism. Stability was achieved without force or defense. Each step required less imagination, not more.

This is precisely where the risk begins.

When an idea explains too much too cleanly, it creates pressure—not to refine understanding, but to formalize it aggressively. Intuition that works invites compression. Geometry that clarifies invites symbolization. And once symbols arrive, they do not merely record insight; they begin to replace it.

Faraday stopped at the edge of that transformation by instinct.

He understood something that later generations would struggle to remember: that explanation must never outrun observation, and that language must remain subordinate to behavior. He was content to see structure without naming its contents, to describe response without appointing agents, to accept completion without dramatizing cause.

But physics did not stop with Faraday.

It moved forward—brilliantly, productively, and with undeniable success. Equations multiplied. Predictive power exploded. Authority consolidated. And somewhere in that success, a quiet shift occurred: intuition was no longer guiding formalism; formalism began guiding intuition.

What had been visual aids hardened into entities. What had been summaries became causes. What had been metaphors acquired ontological weight.

This interlude exists to mark that turning point—not as a failure, but as a drift.

Nothing that follows should be read as an attack on mathematics or a rejection of Maxwell's achievement. What drifted was not correctness, but explanatory discipline. The ability to calculate outpaced the willingness to ask whether the calculation still corresponded to a mechanism—or merely to a symbol that worked.

Faraday's legacy did not disappear. It was absorbed—then transformed.

And that transformation brought extraordinary gains at a quiet cost: the gradual replacement of *how systems behave* with *what symbols describe.* The language of fields, forces, and flux began to sound like ontology rather than bookkeeping. Metaphors became objects. Convenience became explanation.

Next part enters this terrain carefully.

It does not ask, *Where did physics go wrong?* It asks a sharper question: *Where did explanation stop being checked against structure?*

The next chapters examine how a triumph of synthesis subtly changed the rules of understanding—how geometry became algebra, how intuition became abstraction, and how language began to do work that mechanisms once did.

Faraday gave physics clarity without excess. What followed gave physics power—with drift.

To see that drift is not to undo progress. It is to recover a choice we forgot we were making.

And with that awareness restored, we are ready to step into the most delicate part of the story: not the loss of truth, but the quiet substitution of meaning by language.

Part IV — Where Physics Drifted

What was meant as a way of seeing became a thing to believe.
Structure was mistaken for substance, and clarity quietly slipped.

Chapter 9 — Maxwell's Triumph and the Ontology Shift

From intuitive geometry to abstract entities

The Moment Electromagnetism Became Complete

There are rare moments in physics when confusion does not dissolve slowly, but collapses all at once. The late nineteenth century witnessed one such moment.

With the formulation of his equations, James Clerk Maxwell did not merely refine electromagnetism—he completed it. Phenomena that had lived separately for centuries suddenly aligned. Electricity, magnetism, and light fell into a single, coherent mathematical structure. What Faraday had seen intuitively in space, Maxwell expressed with precision across symbols.

The effect was immediate and profound.

For the first time, electromagnetism could predict behavior far beyond the laboratory. Waves could be calculated before they were observed. Propagation could be timed, engineered, and exploited. The equations did not just explain known experiments; they announced new realities. Few achievements in science rival this level of unification.

This was a genuine triumph.

It is important to say this without reservation. Maxwell did not distort Faraday's insight—he honored it. He took Faraday's geometric intuition and gave it reach, durability, and portability.

Ideas that once required careful visualization could now travel intact through textbooks, lectures, and generations. Understanding could be compressed, transmitted, and scaled.

Physics became powerful in a new way.

And power changes habits.

When equations work this well, they begin to feel self-sufficient. When predictions succeed repeatedly, the symbols that produce them acquire authority—not just as tools, but as explanations. The question subtly shifts from *what is happening* to *what the equations say is happening*. Not because anyone decides so, but because success makes resistance feel unnecessary.

This is the moment where the story turns.

Nothing went wrong here. No mistake was made. No fraud entered physics. But something changed in the culture of explanation. Geometry gave way to algebra. Visualization became optional. And the language surrounding fields began to sound less like description and more like declaration.

The equations were flawless. The interpretation quietly loosened its discipline.

This chapter begins at that hinge—not to diminish Maxwell's accomplishment, but to locate its unintended consequence. Faraday gave physics a way to see structure without substance. Maxwell gave physics the power to calculate that structure perfectly.

What followed was not error, but drift.

And drift, when it happens inside success, is the hardest kind to notice.

9.1 — What Maxwell Inherited from Faraday

Before a single equation was written, the conceptual groundwork of electromagnetism had already been laid—quietly, visually, and without ontology—by Faraday. Maxwell did not invent a new picture of nature; he recognized the power of an existing one and set out to preserve it across a different medium.

What Maxwell inherited was not a collection of experimental facts. It was a way of thinking.

Faraday's central contribution was geometric intuition. Influence was not an action at a distance between objects, but a distributed arrangement in space. Lines of force were never substances; they were visual summaries of how systems were poised to respond. Configuration mattered more than components. Boundaries mattered more than forces. Change mattered more than presence.

Maxwell understood this deeply.

He saw that Faraday's drawings were not naïve sketches but structural maps—early attempts to encode spatial relationships that mathematics had not yet learned to express. Faraday had already abandoned point-based causality. He had already moved physics away from localized actors toward field-like continuity, without ever claiming that a "field" existed as a thing.

This inheritance shaped Maxwell's goal.

He did not set out to replace Faraday's intuition. He set out to translate it—to give it a language capable of precision, generalization, and transport across minds that could not see what Faraday saw so naturally. Where Faraday relied on spatial imagination, Maxwell relied on symbolic consistency. But the source insight remained the same.

Crucially, what Maxwell inherited was also Faraday's restraint.

Faraday never asked what lines were made of. He never treated them as carriers or agents. He treated them as relations under constraint. Maxwell respected this stance initially. His early writings are careful. He speaks of models, analogies, and representations—not of ontological commitments. The mathematics is explicitly presented as a way to *represent behavior*, not to declare substance.

At this stage, there was no ontology shift.

Fields were not things. Equations were not causes. Symbols were not reality.

Maxwell's achievement, in this light, was an act of preservation under pressure. He faced a scientific culture that demanded rigor, reproducibility, and calculability. To keep Faraday's insights alive in that environment, he had to encode geometry into algebra. That act was necessary—and brilliant.

But translation always carries risk.

When intuition is converted into symbols, the symbols can outlive the intuition that gave them meaning. Maxwell inherited Faraday's clarity, but he also inherited the burden of ensuring that clarity survived abstraction. As long as the equations were read as *maps of structure*, Faraday's vision remained intact.

The shift came later—not from Maxwell's intent, but from how his success was received.

At the point of inheritance, however, the lineage is clean and unmistakable: Maxwell did not overthrow Faraday. He carried him forward, as faithfully as the tools of his time allowed.

Understanding this continuity is essential, because it shows that the ontology shift was not born of error—but of triumph translated too well.

9.2 — Why Mathematics Was Necessary

The move from Faraday's visual intuition to Maxwell's equations was not a philosophical choice—it was a practical inevitability.

Faraday's way of seeing was powerful, but it was also person-specific. It relied on spatial imagination, patience with geometry, and an ability to think in configurations rather than symbols. That kind of understanding does not scale easily. It cannot be transmitted reliably across classrooms, cultures, or generations without loss. Physics, by its nature, demands a medium that survives translation between minds.

Mathematics provided that medium.

As electromagnetism expanded beyond tabletop experiments into telegraphy, optics, and engineering, the need for precision, generality, and prediction became unavoidable. Visual intuition could show *how* systems were arranged, but it could not tell exactly *how much*, *how fast*, or *how far*. Nor could it guarantee consistency across increasingly complex situations.

Equations could.

Maxwell recognized that without formalism, Faraday's insights would remain powerful but fragile—clear to those who could see them, inaccessible to those who could not. Mathematics offered compression: a way to condense a vast range of spatial behaviors into a small set of relations that could be manipulated, tested, and extended.

This compression was not reduction. It was preservation under constraint.

Importantly, mathematics did not enter as an ontological claim. Maxwell did not say, "This equation is what exists." He said, in effect, "This equation captures how the system behaves." The symbols were intended as representations of relations, not declarations of substance. They were maps, not territories.

Without this step, electromagnetism could not have matured.

Predictive success—such as anticipating electromagnetic waves—required formal structure. Engineering applications demanded calculable quantities. Communication technologies required reliability that intuition alone could not guarantee. Mathematics was the only language capable of carrying Faraday's geometric insights into these domains without distortion.

So the formalization itself was not the problem.

The necessity was real. The achievement was extraordinary. The gain was irreversible.

What matters, however, is understanding what mathematics was meant to do. It was meant to encode structure, not replace it. It was meant to summarize behavior, not redefine causality. As long as equations were treated as disciplined shorthand for Faraday's intuition, understanding remained intact.

The shift did not occur because mathematics was introduced. It occurred because mathematics later came to be treated as self-explanatory.

At this stage—where Maxwell stood—formalism was still humble. It served insight. It followed geometry. It respected the fact that symbols describe relationships, not mechanisms.

Recognizing this is essential. It prevents a false narrative in which abstraction is blamed for drift. Abstraction was necessary. The drift came later—when abstraction was mistaken for explanation itself.

Maxwell's use of mathematics was not the beginning of misunderstanding. It was the price of success—and the condition that made later misunderstanding possible.

9.3 — The Translation That Changed the Rules

The decisive shift did not occur when Maxwell wrote equations. It occurred when geometry was translated into symbols, and symbols quietly became the primary carriers of meaning.

Faraday's intuition lived in space. His understanding depended on seeing how configurations arranged themselves, how boundaries shaped response, and how redistribution unfolded across regions. Meaning was inseparable from spatial relations. When Maxwell translated this intuition, he did so faithfully—but translation always changes what travels best.

In the new language, lines of force became vectors, continuous arrangements became differential relations, and visual density became magnitude. What had been grasped as structure was now expressed as algebraic dependency. This was an extraordinary gain in precision—but it altered how understanding was *experienced*.

Geometry invites insight. Algebra invites manipulation.

Once relations were encoded symbolically, they could be combined, transformed, and generalized without the need to visualize the underlying configuration. This is where the rules of engagement changed. One no longer needed to *see* the system to work with it. One needed only to follow the symbols.

This was not a loss of truth. It was a loss of immediacy.

The translation also altered pedagogy. Students no longer learned electromagnetism by tracing structure; they learned it by mastering notation. Understanding became something that could be postponed—or even bypassed—so long as the equations were applied correctly. Geometry became optional. Intuition became a supplement.

Over time, this produced a subtle inversion.

Originally, symbols were answers to questions raised by structure. After translation, structure was inferred from symbols. What could be written down came to define what was considered real. The equations no longer summarized understanding; they *licensed* it.

This is where ontology began to drift.

Not because Maxwell declared it, but because symbols are persuasive. When an equation works across every experiment, it begins to sound like a statement about what exists rather than a summary of how things relate. Vectors "in space" feel like inhabitants of space. Fields described everywhere begin to feel like things that are everywhere.

The translation succeeded too well.

Maxwell did not insist on this interpretation. In his own writing, he remained cautious, often speaking in terms of models and representations. But once the equations proved universally reliable, the temptation to treat them as *ontological commitments* became hard to resist.

The rules had changed because the medium had changed.

Geometry disciplines imagination by keeping it tied to space and constraint. Algebra disciplines calculation but leaves imagination free to wander—or to disappear entirely. When the latter becomes dominant, explanation begins to drift from mechanism toward symbolism.

This section marks that pivot.

Not the moment of error. Not the moment of betrayal. But the moment when understanding learned to travel without intuition—and in doing so, began to forget where it came from.

9.4 — When Description Began to Sound Like Ontology

At a certain point, the language used to *describe* electromagnetism began to quietly declare what it was.

This shift did not happen through argument or proclamation. It happened through repetition, convenience, and success. As Maxwell's equations proved universally reliable, the terms embedded within them—*field*, *energy density*, *propagation*—started to feel less like bookkeeping and more like statements about what exists in the world.

The distinction between *representation* and *reality* blurred.

Originally, a field was a way to summarize how space was configured to respond. It was a descriptive scaffold—a compact way to say, "if you place something here, this is how the system will behave." But as equations became the primary interface with the phenomenon, the field began to sound like an entity occupying space, capable of acting, storing, and transmitting.

Nothing in the mathematics demanded this interpretation. The language encouraged it.

Phrases such as "the field carries energy," "the field propagates," or "the field does work" were shorthand for well-defined outcomes. Over time, the shorthand hardened. What began as a convenient compression of behavior started to read like ontological attribution. The map was no longer clearly labeled as a map.

This is where explanation subtly changed character.

Instead of asking how a configuration redistributes imbalance, explanations began to reference what the field "does." Instead of tracing boundary conditions and geometry, discourse leaned on field behavior as if it were self-explanatory. The field became an answer, not a summary.

Importantly, this was not a mistake introduced by Maxwell himself. Maxwell was careful, often explicit, about the representational status of his constructs. The drift occurred later—when his equations were inherited without his caution, and when pedagogical efficiency favored nouns over mechanisms.

Ontology arrived by implication, not by proof.

Once fields were treated as things, further questions became unnecessary. If a field exists everywhere, then influence everywhere is explained. If energy resides in the field, then transfer requires no additional account. The explanatory chain stops—not because it has reached bedrock, but because it has found a comfortable resting place.

This comfort is deceptive.

Descriptive success does not confer ontological truth. A language that predicts outcomes perfectly can still mislead about *why* those outcomes occur. When description sounds like existence, understanding risks becoming linguistically complete but mechanically thin.

This section marks this precise inflection point: where talking *about behavior* quietly turned into talking *about being*— and where physics, without noticing, allowed its metaphors to graduate into reality.

9.5 — The Field Becomes a Thing

Once the language of description began to sound ontological, the next step followed almost automatically: the field stopped being a tool and started behaving like an actor.

At first, this felt harmless—even helpful. Saying "the field does work" or "energy resides in the field" offered a clean way to speak about distributed effects without tracing every boundary and configuration. The shorthand was efficient. Communication improved. Calculations aligned neatly with results.

But efficiency has side effects.

Gradually, the field was no longer treated as a relational summary. It was treated as something that *exists independently*, fills space, carries properties, and performs actions. Influence was no longer something that emerged from configuration; it was something the field *exerted*. Redistribution no longer needed to be traced; it was assumed to be handled by the field itself.

This is the moment where ontology quietly solidified.

The field crossed a conceptual threshold—from descriptive construct to assumed entity. Once that happened, explanations began to close prematurely. If the field is real and active, then asking *how* influence propagates feels redundant. The field becomes the cause, not the summary of causes.

What was lost here was not predictive power. Nothing broke. Equations continued to work flawlessly. What was lost was mechanistic accountability.

Boundaries became less important because the field was everywhere. Geometry mattered less because the field already "knew" how to act. Redistribution became implicit because the field was assumed to manage it.

This shift also changed intuition.

Students and practitioners began to imagine space as filled with something that pushes, pulls, stores, and transmits—rather than as a structured arena where imbalance resolves through allowed pathways. The original Faraday-style question—*what arrangement makes this behavior unavoidable?*—was replaced by a more passive one: *what is the field doing here?*

That question feels explanatory. But it is often terminal.

It is crucial to note that this was not a deliberate philosophical move by James Clerk Maxwell. Maxwell used the language available to him and repeatedly emphasized models and analogies. The reification happened after him—through repetition, pedagogy, and the sheer authority of success.

Once fields became things, they could be stacked, coupled, quantized, and promoted without ever revisiting their original status as summaries of structure. Ontology expanded quietly, not because evidence demanded it, but because language made it comfortable.

This section is not arguing that fields are "wrong." It is arguing that treating them as actors changed what questions physics felt allowed to ask.

When a tool becomes a thing, explanation stops tracing behavior and starts naming entities. And when that happens, understanding does not collapse—it simply stops deepening.

This is the core of the ontology shift: not the introduction of a false object, but the retirement of structural questioning in favor of linguistic certainty.

9.6 — What Was Gained (And Cannot Be Unlearned)

It is essential to pause here and state this without ambiguity: nothing that followed Maxwell was a regression. The ontology shift did not erase understanding; it unlocked a level of capability that physics could never relinquish again.

What was gained was predictive sovereignty.

With Maxwell's formalism, electromagnetism ceased to be a domain of clever experiments and became a fully generative framework. One could calculate outcomes before building apparatus. One could design systems that had never existed and trust they would behave as predicted. This was not philosophical progress—it was operational mastery.

The consequences were irreversible.

Electromagnetic waves were not merely explained; they were *anticipated*. Communication technologies emerged not by trial, but by calculation. Power generation, motors, transmission lines, radio, optics—all of modern infrastructure traces directly to this formal unification. No intuitive geometry alone could have carried physics into this territory.

This is why the shift cannot be undone.

Once equations demonstrate universal reliability, they become indispensable. They are no longer optional viewpoints; they are structural infrastructure for civilization. Any framework that proposes to "go back" misunderstands the nature of the gain. Physics did not choose formalism over intuition—it outgrew the limits of intuition-only reasoning.

Something else was gained as well: collective scalability.

Faraday's insight depended on rare minds capable of sustained spatial reasoning. Maxwell's equations allowed average minds to participate productively. Knowledge could be standardized, taught, examined, and applied consistently across continents.

Understanding no longer depended on seeing—it depended on following rules. This democratization of capability was not trivial; it reshaped science itself.

And then there is unification.

For the first time, electricity, magnetism, and light were not cousins—they were expressions of the same structure. This was not merely a conceptual victory; it was a compression of reality. Fewer principles explained more phenomena. That achievement alone places James Clerk Maxwell among the central architects of modern physics.

Nothing in this book disputes that.

The key point of this section is not to diminish what was gained, but to separate gain from interpretation. Power increased dramatically. Reliability became absolute. Technology flourished. None of these outcomes require fields to be ontological actors—but once formalism delivered them, the temptation to treat it as complete explanation became overwhelming.

This is why the shift persists.

You cannot ask a civilization to abandon the language that built it. You cannot unlearn equations that work everywhere. You cannot pretend that intuition alone could replace a framework that produces satellites, lasers, and global networks.

What was gained is permanent.

But permanence of use does not imply permanence of interpretation.

This distinction matters, because it allows the next step to be taken honestly. One can accept every achievement of Maxwellian physics—every equation, every prediction, every technology—without freezing its metaphors into ontology.

This section establishes this ground clearly:

- The triumph was real

- The gains were decisive
- The success was deserved
- And none of it needs to be undone

What follows in the chapter is not a rejection of this success, but an examination of its unintended cognitive side effects—what quietly slipped away once explanation felt finished.

Physics did not lose truth here. It gained so much power that it stopped checking how it was explaining what it already knew.

And that is a very different kind of problem.

9.7 — What Quietly Slipped Away

The gains of Maxwellian formalism were obvious, measurable, and celebrated. What slipped away did not announce itself with failure. It faded silently, precisely because nothing seemed broken.

The first thing that receded was mechanism.

As equations grew more powerful, the need to ask *how* a phenomenon unfolded in space weakened. If the symbols predicted outcomes correctly, tracing the actual redistribution of energy, the role of boundaries, or the sequence of structural constraints began to feel optional. Explanation shifted from *process* to *permission*: if the equations allow it, it must be happening.

Mechanism was not disproved. It was bypassed.

The second loss was structural accountability.

Faraday's questions were always geometric: What arrangement makes this behavior unavoidable? Where does imbalance resolve? Which paths are allowed, and which are closed? These questions keep explanation tethered to physical structure. Once the field became an assumed actor, those questions lost urgency. The field "handled" them implicitly.

Boundaries became background. Configuration became context. Redistribution became assumed.

The third thing that slipped away was epistemic humility.

When explanation rests on symbols that work everywhere, it becomes tempting to believe that understanding is complete. Ambiguity feels resolved, not because it has been addressed, but because it has been rendered unnecessary. The equation answers faster than curiosity can ask.

This does not produce error. It produces premature closure.

A subtler loss followed: the discipline of interpretation.

Earlier thinkers were forced to constantly distinguish between description and cause, between model and reality. With success, that distinction weakened. The language of fields, energy density, and propagation began to sound less like shorthand and more like substance. Interpretation hardened without being examined.

This is not because James Clerk Maxwell intended it. It is because later generations inherited his equations without inheriting his caution. What began as representational convenience quietly became explanatory finality.

Finally, something deeply human slipped away: the habit of looking.

When calculation dominates, observation becomes confirmatory rather than exploratory. Experiments are designed to validate equations, not to challenge interpretation. Anomalies are treated as edge cases or noise, not invitations to rethink structure. Nature is consulted less often—not because it has nothing to say, but because symbols already seem to speak for it.

None of this damaged physics operationally. That is precisely why it persisted.

What quietly slipped away was not correctness, but depth. Not capability, but curiosity disciplined by structure. The kind of understanding that refuses to stop at "it works" and insists on asking "what makes it unavoidable?"

This section marks that loss clearly—not to mourn it, but to make it visible again.

Because once you see what slipped away, you realize something important: it was never incompatible with formalism. It was simply no longer required.

And what is not required is rarely maintained—no matter how valuable it once was.

9.8 — The Ontology Shift in Retrospect

Seen from a distance, the ontology shift looks less like a decision and more like a momentum effect.

No committee voted to declare fields real entities. No philosopher issued a decree. No experiment forced the conclusion.

What happened instead was cumulative inheritance.

Each generation received a framework that worked extraordinarily well and asked fewer questions about *why it worked the way it did*. The language used to teach it was slightly more concrete than the language used to create it. Metaphors hardened. Shortcuts multiplied. And slowly, what had begun as representational convenience settled into assumed reality.

In retrospect, this was almost inevitable.

When a theory succeeds across every scale available to testing, the distinction between *model* and *world* begins to feel pedantic. If treating fields as things produces no immediate contradiction—and delivers technology, prediction, and coherence—why resist that interpretation? Ontology becomes the path of least resistance.

This is why it is important to say clearly: James Clerk Maxwell is not the culprit here.

Maxwell stands at a hinge point, not a fault line. He completed the mathematical unification of electromagnetism without insisting that the mathematics define existence. The drift occurred after him, through pedagogical simplification, institutionalization, and the natural human desire for nouns that feel solid.

What later generations inherited was not Maxwell's careful posture, but Maxwell's success.

Once equations became the unquestioned foundation of explanation, interpretation quietly inverted. Instead of asking whether language still corresponded to mechanism, physics asked only whether predictions remained accurate. Ontology was allowed to ride along unexamined, because it did not interfere with results.

Over time, this produced a subtle but consequential change in worldview.

Fields were no longer seen as ways of describing spatial readiness. They were seen as occupants of space. Energy was no longer a bookkeeping of redistribution. It was seen as something carried and stored by those occupants.

Nothing exploded. Nothing failed. But explanation became ontologically heavier than necessary.

In retrospect, the shift is best understood not as an error, but as a trade-off accepted unconsciously. Physics gained extraordinary reach and lost a degree of interpretive restraint. It learned to calculate perfectly while asking fewer questions about structural inevitability.

Recognizing this does not weaken Maxwellian physics. It liberates it from unnecessary metaphysics.

This retrospective view allows us to separate three things that were quietly conflated:

- Mathematical correctness
- Explanatory sufficiency
- Ontological commitment

Maxwell guaranteed the first. History assumed the second implied the third.

This section closes the chapter by restoring that distinction. It shows that the ontology shift was not demanded by nature, but permitted by success. And what is permitted without resistance often becomes invisible.

With that clarity in place, the path forward becomes possible—not by undoing Maxwell, but by recovering Faraday's restraint inside Maxwell's power.

That recovery begins next, where language itself comes under examination— and where metaphors, stretched too far, begin to replace mechanisms altogether.

Closing Transition — From Completion to Drift

By the end of this chapter, the landscape is clear.

Electromagnetism did not drift because it was misunderstood. It drifted because it was understood too well, too quickly, and too successfully.

James Clerk Maxwell closed a chapter that Faraday had opened with extraordinary care. He gave physics a system that could predict, unify, and scale beyond anything intuition alone could manage. That achievement stands untouched. Nothing in what follows diminishes it.

But completion changes behavior.

Once a framework feels finished, curiosity changes its posture. Questions that once mattered become optional. Mechanisms that once demanded explanation are quietly assumed. Language, no longer checked against structure, begins to stand in for it. What was once a map becomes comfortable enough to mistake for terrain.

This chapter marks that inflection point precisely.

Not where physics failed. Not where mathematics overreached. But where interpretation stopped being policed by restraint.

Maxwell completed the equations. Physics inherited the power. And in that inheritance, a subtle substitution occurred: symbols began to answer questions they were never meant to ask.

What follows next is not a continuation of abstraction, but an examination of its side effects. Next chapter turns directly to language itself—to words like *field*, *force*, and *flux*—and asks how metaphors, stretched beyond their original intent, began to function as explanations rather than aids.

If this chapter showed *where* the drift began, next chapter will show *how* it was sustained.

The story now moves from equations to vocabulary, from symbols to speech, from formal success to explanatory comfort.

And only by understanding that transition can we begin the larger task ahead: recovering clarity without surrendering power.

With that, the stage is set.

The equations remain. The success is intact. But the language is about to be questioned.

Chapter 10 — When Language Replaced Mechanism

Field, force, and flux as metaphors overstretched

The Comfort of Saying Something

There is a subtle relief that comes with having the right words. When a phenomenon resists intuition, language steps in to steady the mind. To say *field*, *force*, or *flux* feels like progress. The words are familiar, authoritative, and repeatable. They give the impression that something definite has been identified—even when no mechanism has been traced.

This is not deception. It is comfort.

In physics, language often arrives after success. Once equations predict reliably and experiments align cleanly, vocabulary crystallizes around those results. The words do not arise to explain *how* something happens, but to make it easier to talk about *that it happens*. Over time, this distinction fades. The act of naming begins to feel like the act of explaining.

Confidence settles in quietly.

A student hears that a charge moves "because of the field." An engineer learns that energy flows "through flux." A textbook states that forces "act at a distance."

Each sentence sounds complete. None of them describe a process.

The danger here is not incorrectness. The danger is premature closure. When a term becomes familiar enough, it stops inviting questions. It begins to function as an answer rather than a pointer. Curiosity dissolves not through refutation, but through satisfaction.

This is where explanation drifts—not mathematically, but linguistically.

Words that were introduced to bridge understanding gradually replace the bridge itself. They stop referring back to structure, configuration, and redistribution. Instead, they acquire an independent authority. To name is to conclude. To repeat is to reinforce.

What follows in this chapter is not a critique of terminology, but an examination of how terminology quietly took on explanatory weight it was never meant to carry. The equations still stand. The predictions still hold. But the language wrapped around them has grown louder than the mechanisms beneath.

This chapter begins with a simple claim: that physics did not lose rigor when language expanded— it lost discipline in how language was used.

To recover understanding, we must first notice where words began to sound like causes, and why that sounded convincing enough to let us stop looking.

10.1 — Why Physics Needed Metaphors

Physics did not begin with equations. It began with human perception struggling to keep up with unfamiliar behavior.

Electricity, magnetism, and later electromagnetic propagation were never directly visible. Their effects could be felt, measured, and recorded—but the processes themselves unfolded beyond ordinary sensory access. In such situations, language had to do what perception could not: build a bridge between experience and abstraction.

Metaphors were that bridge.

Words like *force*, *field*, and *flux* entered physics not as declarations of reality, but as cognitive scaffolding. They gave thinkers something to hold onto while navigating phenomena that resisted everyday intuition. A "force" echoed muscular effort. A "field" echoed agriculture and space-filling presence. "Flux" echoed flow and movement. Each term borrowed familiarity to soften conceptual shock.

This was not a weakness. It was a necessity.

Without metaphors, early understanding would have stalled. Metaphors allowed physicists to speak before they could see clearly, to reason before mechanisms were fully articulated. They acted as placeholders—temporary supports that enabled communication, collaboration, and incremental progress.

Crucially, in their original role, metaphors were directional, not terminal. They pointed attention toward patterns of behavior. They invited refinement. They were meant to be outgrown.

In Faraday's hands, this discipline was clear. His language was cautious. "Lines of force" were explicitly visual aids, not threads in space. They were ways of tracking how systems tended to respond, not claims about what filled the void. The metaphor served observation; it did not replace it.

The same was true in early formalism. Words accompanied equations as interpretive aids, not as causal agents. They helped explain results to human minds without pretending to be the mechanism themselves.

The problem did not begin with metaphor.

The problem began when metaphors stopped behaving like bridges and started behaving like foundations.

When a metaphor is treated as provisional, it sharpens understanding. When it is treated as sufficient, it freezes it.

This section establishes this clearly: physics needed metaphors to move forward at all. Without them, abstraction would have been inaccessible and progress impossible. The drift explored in this chapter arises not from their introduction, but from their unexamined permanence.

To understand where language later replaced mechanism, one must first acknowledge why language was invited in—and how carefully it was originally meant to behave.

10.2 — From Helper to Placeholder

Metaphors begin as aids, but they do not remain neutral.

Once a term proves useful, it gets reused. Once it gets reused, it becomes familiar. And once it becomes familiar, it starts to stand in for the work it was meant to support. This is the quiet transition from helper to placeholder.

In physics, this transition is especially easy to miss because nothing appears to break.

A metaphor like *field* initially helps organize observations: it groups responses, summarizes tendencies, and compresses spatial behavior into a manageable idea. But when the underlying mechanism remains difficult to articulate—and the predictions keep succeeding—the metaphor stops pointing backward to structure. It begins pointing only to itself.

At that moment, language fills a gap.

Instead of saying, "We do not yet know how this redistribution completes," one says, "The field does it." Instead of tracing geometry and boundary conditions, one says, "Because of flux." Instead of asking why a configuration must respond this way, one says, "That's the force."

Each substitution feels harmless. Each one saves time. Each one preserves confidence.

But cumulatively, something changes: the question is no longer carried forward.

The metaphor becomes a verbal resting place. It signals that an explanation exists—even if no mechanism has been described. The listener feels informed. The speaker feels precise. Inquiry pauses, not because it has been satisfied, but because it has been linguistically deferred.

This is how placeholders work.

They occupy the space where a mechanism should be, without announcing that the space is still empty. They allow discourse to continue smoothly while postponing deeper tracing indefinitely. Over time, the placeholder is no longer recognized as temporary. It becomes part of the conceptual furniture.

What makes this transition especially resilient is success.

Because the equations work, the placeholder is never challenged by failure. Because experiments align, the language is never pressured to refine itself. The metaphor does not need to earn its keep; it has already been paid for by predictive accuracy.

At this stage, explanation has not become false. It has become sufficient by default.

This section identifies this moment precisely. It is not the misuse of language, but its overextension. A tool designed to assist understanding quietly takes over the role of understanding itself. And once that happens, physics continues forward—confident, productive, and increasingly detached from the mechanisms its words were once meant to illuminate.

From here on, the drift is no longer accidental. It is self-reinforcing.

Because every time language answers a question prematurely, it trains the next generation to stop asking it.

10.3 — "Field" as a Word That Grew Too Large

No word illustrates the linguistic drift in physics more clearly than *field.*

Originally, the field was a relational convenience. It was a way of saying: *given this configuration, this is how space is prepared to respond.* It summarized tendencies. It organized expectations. It allowed physicists to speak compactly about distributed behavior without tracking every interaction explicitly.

At this stage, the field was not a thing. It was a description of readiness.

But the word had a dangerous advantage: it sounded spatial. A field felt like something that could exist "everywhere," even when it was only describing how a system would behave *if disturbed.* Over time, this linguistic texture began to do conceptual work that the original idea never intended.

The field started to answer questions by existing.

Why does a charge accelerate here? Because there is a field here. Why does influence propagate through empty space? Because the field fills it. Why does energy appear distributed? Because the field carries it.

Each answer feels explanatory. None of them trace a mechanism.

As the field grew conceptually larger, it also grew less discriminating. Boundaries mattered less, because the field was already everywhere. Geometry mattered less, because the field encoded response automatically. Configuration-specific inevitability—Faraday's core insight—was replaced by a universal backdrop that "handles" behavior globally.

This shift subtly changed how problems were framed.

Instead of asking, *what arrangement makes this response unavoidable?* Physics began asking, *what is the field doing here?*

That question feels natural. It is also terminal.

Once the field is treated as an entity, further tracing becomes unnecessary. Redistribution does not need to be followed. Completion does not need to be located. The field has already been granted agency, and agency ends inquiry.

It is important to see that this expansion of the field was not demanded by mathematics. The equations describe relations between quantities defined at points and regions; they do not assert that a substance occupies space and acts. The enlargement occurred in interpretation, not formalism.

Language did the work.

As textbooks, lectures, and explanations repeated the term without reattaching it to its original descriptive role, the field quietly crossed a threshold. It stopped being a map of structure and became an inhabitant of space. Once that happened, it became very hard to ask structural questions without sounding naïve.

Why worry about boundaries when the field is everywhere? Why track redistribution when the field already "stores energy"? Why look for completion when the field propagates endlessly?

These are not errors. They are consequences of an overgrown word.

This section makes one point sharply: the problem is not that fields are used, but that they are used as explanations rather than summaries. When a word grows large enough to replace mechanism, it no longer clarifies—it concludes.

And when a concept concludes too easily, understanding stops not because it is complete, but because it feels that way.

10.4 — "Force" as an Inherited Reflex

If *field* grew too large, *force* simply refused to leave.

Force is one of the oldest words in physics, and it carries a deep psychological weight. It resonates with bodily experience—push, pull, effort, resistance. Long before electromagnetism, force provided a reliable way to think about motion. When something moved, something must have *acted* on it. The logic felt universal.

That reflex survived the transition into electromagnetism almost unchanged.

Even when phenomena no longer resembled push–pull interactions, force language remained the default explanatory instinct. Charges were said to experience forces. Currents were deflected by forces. Motion was attributed to forces—even when no single direction, source, or point of application could be identified.

The language persisted because it felt familiar.

But electromagnetism does not behave like mechanics.

In many electromagnetic situations, nothing is pushed. No agent applies effort. Motion emerges not because something *acts*, but because a configuration cannot resolve imbalance in any other way. Geometry, constraints, and redistribution do the work that force language pretends to describe.

Force becomes a label, not a cause.

This is especially clear in systems like the electric motor, induction, or the Faraday disk—where motion is continuous, direction is emergent, and no isolated force can be cleanly identified. Yet the reflex remains: if something moves, there must be a force responsible.

So the word stays, even when it no longer earns its place.

What makes force language misleading here is not that it is wrong, but that it is structurally lazy. It bypasses the need to ask *why this configuration must evolve this way*. It compresses a distributed, boundary-dependent process into a single imagined action.

Force answers too quickly.

Instead of tracing how imbalance redistributes across space, force language assigns causality to an abstract push. Instead of examining constraints, it assumes an agent. The explanation feels intuitive, but it disconnects understanding from structure.

This is why force language persists so stubbornly.

It satisfies the mind's desire for agency. It mirrors everyday experience. And it allows explanation to end without discomfort.

But electromagnetism does not need force to operate. It needs asymmetry that cannot resolve statically. When that asymmetry exists, motion follows—not because something pushes, but because stillness is no longer an option.

This section highlights this inheritance clearly: force remained in electromagnetic language not because it explained more, but because it explained easily. It survived as a reflex, not as a necessity.

And once force becomes reflexive, it becomes invisible.

Physics continues to speak of forces long after structure has replaced them— and in doing so, language quietly pulls explanation backward, toward intuition that no longer matches the phenomenon.

This is not a failure of physics. It is a reminder that words, once useful, can outlive their usefulness— and still feel indispensable.

10.5 — "Flux" as Movement Without Mechanism

Among all the terms that migrated from description to explanation, *flux* may be the most deceptive—because it sounds dynamic without ever specifying what actually moves.

Flux entered physics as an accounting device. It was never meant to be an actor. It answered a narrow, technical need: to track how much of something—field lines, influence, or response tendency—passed through a surface as conditions changed. It was a bookkeeping concept, not a causal one.

At first, this discipline was clear.

Flux did not *do* anything. It recorded change. It summarized redistribution.

But as with *field* and *force*, repeated use slowly altered its role.

When induction was formalized, statements like "changing flux induces current" became standard. Mathematically, this is correct. Pedagogically, it is efficient. Conceptually, it is dangerous. Because the sentence subtly implies that flux is an agent—that something *moves*, gets *cut*, or *acts*.

Flux began to sound like motion.

Once that happened, explanation quietly inverted. Instead of asking how a configuration fails to remain in equilibrium and therefore redistributes, one says: the flux changes, therefore current appears. The causal arrow feels complete. Nothing more seems required.

But flux has no mechanism.

It does not specify *where* redistribution occurs. It does not identify *how* imbalance propagates. It does not explain *why* the system cannot settle statically.

It merely tallies an outcome.

The problem deepens when metaphors accumulate. "Flux cutting," "flux linkage," "flux flow"—each phrase borrows physical intuition without supplying physical process. The words suggest motion, but no substance moves. They suggest interaction, but no interaction is traced. They suggest causality, but no constraints are identified.

What results is a narrative shortcut.

Instead of following energy redistribution through geometry and boundaries, flux becomes a story we tell ourselves about change. It feels explanatory because it tracks a variable that changes when something happens. But correlation is not mechanism—and flux, by design, was never meant to be more than a correlational measure.

This is why flux-based explanations struggle in edge cases.

Devices like the Faraday disk, rotating conductors, or non-standard geometries expose the weakness immediately. The mathematics still works. The language breaks down. "Flux cutting" no longer maps cleanly to behavior, and explanation retreats into technical phrasing without clarity.

Flux did not fail. It was asked to do a job it was never designed for.

This section isolates this mistake precisely. Flux is invaluable as a diagnostic and computational tool. But when it is promoted from measurement to cause, understanding becomes circular: current appears because flux changes, and flux changes because current appears.

That loop feels closed. It is not explanatory.

In structural terms, flux does not drive systems. Systems drive flux values. Redistribution produces flux variation, not the other way around. Reversing that order does not break equations—but it empties explanation of content.

This is the central warning of the section: when bookkeeping terms begin to narrate events, physics risks mistaking its ledgers for its engines.

Flux was never an engine. It was a meter.

And meters, no matter how precise, do not make things happen.

10.6 — When Words Start Ending Questions

The most serious consequence of linguistic drift is not confusion—it is finality.

When words begin to function as explanations, they quietly acquire the power to end inquiry. A term that once pointed toward a mechanism now replaces the need to find one. The conversation stops not because understanding has been reached, but because a familiar word has been spoken.

This is how explanation becomes verbal.

A student asks why a charge accelerates. "The field causes it."

An experiment behaves unexpectedly. "It's a force effect."

A current appears in a loop. "Because the flux changed."

Each response is technically acceptable. Each one is incomplete. And each one feels sufficient enough to discourage further questioning. The word stands in the place where tracing should continue.

What makes this especially insidious is that nothing contradicts the equations. Predictions remain accurate. Calculations remain consistent. There is no immediate pressure to revisit interpretation. The vocabulary performs well socially and pedagogically, even as it displaces mechanism.

Language becomes a soft endpoint.

Once this pattern sets in, curiosity is retrained. Instead of asking *how does this configuration force redistribution?*, one learns to ask *what term applies here?* Understanding becomes categorical rather than structural. Recognition replaces reasoning.

This shift also changes how anomalies are handled.

When an experiment does not fit neatly into existing language, it is treated as a special case rather than a prompt to revisit structure. New terms are introduced. Old ones are stretched. The vocabulary grows thicker, but explanation grows thinner. The system remains operationally intact, but conceptually layered.

Words accumulate where mechanisms should clarify.

The deeper issue is psychological. Familiar language reduces cognitive tension. It provides closure without effort. To say "the field does it" is easier than to trace redistribution through geometry and boundaries. Over time, ease becomes expectation. Anything that demands more feels unnecessary or even unscientific.

This is how language begins to police curiosity.

Questions that push beyond vocabulary sound naïve. Requests for mechanism feel like regressions. The discipline quietly shifts from understanding phenomena to maintaining terminological fluency. One learns how to speak physics before learning how to see it.

This section names this condition precisely: when words stop pointing and start concluding, understanding stalls. Not because nature has nothing more to reveal, but because language has claimed the final word.

The tragedy here is subtle.

Physics does not lose its power. It loses its patience.

And once patience is gone, explanation becomes a matter of recitation rather than exploration—efficient, correct, and increasingly detached from the structures that first made it meaningful.

10.7 — Teaching, Textbooks, and the Reinforcement Loop

Once language hardens into explanation, education becomes its most efficient amplifier.

Textbooks do not drift maliciously; they optimize for transmission. Concepts that can be named are easier to test than processes that must be traced. Nouns travel better than mechanisms. As curricula compress, vocabulary rises to the surface while structural reasoning sinks quietly underneath.

This creates a reinforcement loop.

Students are rewarded for recognizing terms—*field*, *force*, *flux*—and for deploying them fluently in problem-solving contexts. Exams validate substitution, not tracing. Correct answers arrive through symbol manipulation, not through explaining why a configuration cannot resolve any other way. Over time, success becomes decoupled from understanding.

Teachers inherit this structure.

Pressed for coverage and clarity, instruction leans on established language. Metaphors are repeated because they are already familiar. Caveats are dropped because they complicate delivery. What began as shorthand becomes curriculum. What began as a bridge becomes the road itself.

And the loop closes.

Students who succeed within this system become the next generation of instructors, authors, and practitioners. They pass on what worked for them: the same vocabulary, the same explanatory shortcuts, the same comfort with terms that conclude rather than illuminate. The drift is not only preserved—it is institutionalized.

Crucially, nothing in this loop triggers correction.

Predictions remain accurate. Engineering outcomes succeed. Research progresses. There is no penalty for linguistic substitution because operational performance does not depend on interpretive discipline. The system rewards fluency and delivers results. From the outside, it looks like health.

From the inside, depth thins.

This is why the drift is so durable. It is not sustained by ignorance, but by competence. The language works well enough to carry physics forward, and so its limitations remain invisible. To question it can even feel disruptive—like reopening settled matters.

But what settles is not mechanism. It is habit.

This section makes this clear: language drift did not spread because individuals misunderstood physics. It spread because education optimized for efficiency, and efficiency favors words that stand still over explanations that must move.

The result is a discipline that speaks with confidence while quietly forgetting how to look beneath its own vocabulary.

Reversing this does not require new equations. It requires reintroducing a simple demand into teaching and learning:

Do not accept a word as an answer until it points back to a structure that makes the behavior unavoidable.

Without that demand, the loop continues—smoothly, productively, and increasingly detached from the mechanisms it once set out to understand.

10.8 — Why This Drift Felt Harmless

The most revealing aspect of linguistic drift in physics is not that it occurred—but that it never triggered alarm.

Nothing appeared to go wrong.

Predictions continued to match experiments with extraordinary precision. Engineering outcomes improved decade after decade. Technologies based on electromagnetic theory reshaped the world. From radios to satellites, from power grids to computation, the results spoke loudly in favor of the existing framework.

So why worry?

This is precisely why the drift felt harmless: operational success masked explanatory substitution.

When a system delivers reliable outcomes, the human mind relaxes its standards for explanation. If the bridge stands and traffic flows, few people ask whether the blueprint still corresponds to how the structure actually distributes stress. Function becomes evidence of understanding—even when understanding has quietly thinned.

Language benefited enormously from this success.

Words like *field*, *force*, and *flux* never failed in practice. They slotted neatly into equations that worked everywhere they were applied. They enabled problem-solving without demanding interpretive effort. They allowed complex systems to be handled efficiently by people who never had to visualize the underlying structure.

From a practical standpoint, this was a triumph.

From an explanatory standpoint, it created a blind spot.

Because nothing broke, nothing demanded repair. Because no contradiction appeared, no reevaluation felt urgent. Linguistic shortcuts did not produce errors; they produced comfort. And comfort is far more dangerous to inquiry than confusion.

Another reason the drift felt harmless is that it was incremental.

No single metaphor overreach was dramatic enough to stand out. Each step felt reasonable. Each reuse of language felt justified. Over time, the accumulation became invisible. What once would have sounded metaphorical now sounded literal, simply because it had been repeated long enough to feel natural.

Familiarity replaced scrutiny.

There was also a cultural factor at work. Physics increasingly defined rigor through mathematical correctness, not interpretive clarity. As long as equations were sound, explanations were treated as secondary. Language became something to *get through*, not something to discipline carefully. Ontology drifted under the radar because ontology was no longer where rigor was enforced.

In this environment, asking for mechanism began to feel optional—or even unproductive.

Why reopen questions that do not improve predictions? Why challenge language that delivers results? Why complicate explanations when calculations already succeed?

These are reasonable questions. They are also the reason drift persists.

This section makes one point unmistakably clear: the problem was never visible because success hid it. The drift did not weaken physics; it simply detached explanation from structure while leaving performance intact.

That is why this book does not argue that something "went wrong." It argues that something went unnoticed.

And what goes unnoticed in success tends to become permanent—until someone asks a question that language can no longer answer without pointing back to mechanism.

That question is now unavoidable.

And with it, the path opens—not toward rejection of existing theory, but toward reclaiming explanatory discipline without sacrificing power.

10.9 — Reclaiming Language Without Abandoning It

The solution to linguistic drift in physics is not to purge language, nor to invent a new vocabulary wholesale. That impulse simply repeats the same mistake in reverse. The task is more disciplined—and more demanding: to return language to its proper role as a pointer, not a cause.

Words are unavoidable. Human understanding requires them. But words must remain subordinate to structure.

Reclaiming language begins by restoring a simple rule that Faraday instinctively obeyed and later generations quietly relaxed:

No term should be allowed to explain a phenomenon unless it can be traced back to a configuration that makes the behavior unavoidable.

Under this rule, familiar words remain usable—but they lose their authority to conclude inquiry.

- *Field* returns to meaning "how space is configured to respond under constraint," not "what exists everywhere and acts."
- *Force* becomes a shorthand for "the direction in which imbalance resolves," not an invisible push.
- *Flux* resumes its role as a measure of redistribution, not the engine that causes it.

Nothing is removed. Everything is re-anchored.

This shift does not weaken physics; it sharpens it. When language is disciplined by mechanism, explanation regains depth without losing efficiency. Equations remain untouched. Predictions remain intact. What changes is the direction of causality in explanation—from words back to structure.

Under a UED-aligned interpretation, this restoration becomes systematic.

Language is treated as an index, not an ontology. Every term points to:

- configuration
- boundary conditions
- allowed and forbidden redistribution paths
- equilibrium-seeking completion

If a word cannot be translated back into these terms, it is flagged—not rejected, but held as provisional. It may remain useful, but it no longer claims explanatory sufficiency.

This approach also changes pedagogy.

Instead of teaching students to answer questions with terms, it trains them to unpack terms into structures. The question shifts from "what word applies here?" to "what arrangement makes this outcome inevitable?" Language becomes a compression tool for reasoning that has already occurred, not a substitute for it.

Most importantly, reclaiming language restores intellectual humility.

It acknowledges that success in calculation does not equal completeness in understanding. It permits physics to say, "We can predict this perfectly, and we are still refining how we explain it." That statement is not weakness. It is maturity.

This section closes the chapter by drawing a clear boundary:

- Language is necessary
- Metaphors are useful
- Vocabulary is unavoidable

But none of them are causes.

When language is returned to its rightful place—serving mechanism rather than replacing it—physics regains something it never meant to lose: the habit of tracing inevitability back to structure.

With that habit restored, the drift examined in this part of the book does not demand correction through rejection. It invites a more precise continuation—one where power and clarity coexist, and where words once again know what they are standing in for.

That restoration prepares the ground for what follows next: not critique, but reconstruction—with language finally disciplined enough to support it.

Closing Transition — From Words Back to Structure

This chapter closes without overturning a single equation.

Nothing calculated in electromagnetism needs to be undone. The mathematics remains exact. The predictions remain unmatched. The technologies remain triumphant. What has been examined here is not the machinery of physics, but the voice with which it explains itself.

Language did not betray physics. It outgrew its discipline.

Words like *field*, *force*, and *flux* entered the conversation as guides—temporary supports meant to orient intuition while deeper structure was being explored. Over time, their usefulness allowed them to linger. Repetition granted them authority. And eventually, they began answering questions by standing still, rather than pointing onward.

This is the quiet moment when mechanism yields to metaphor.

The drift did not corrupt understanding; it relaxed the demand for it. Explanation became fluent, efficient, and comfortable. And comfort, when paired with success, rarely invites revision.

But Faraday's restraint still offers a way forward.

He did not reject language; he refused to let it replace looking. He did not deny symbols; he kept them tethered to behavior. That discipline is not lost—it has simply been waiting beneath layers of familiarity.

This chapter restores a single, decisive orientation:

Words are summaries of structure, not substitutes for it.

With that orientation reestablished, the story can move beyond critique. The task ahead is not to dismantle Maxwellian physics, nor to abandon its language, but to rebuild interpretation on firmer ground—where terms point back to configuration, boundaries, and completion rather than closing inquiry.

What follows next is the constructive phase.

Having identified where explanation drifted—from mechanism to metaphor, from structure to speech—the book is now ready to propose a framework that keeps language disciplined without sacrificing power.

The equations will remain. The words will remain. But they will be asked to earn their place by pointing back to inevitability.

From here, physics is no longer being corrected. It is being realigned—quietly, precisely, and without drama.

The path now opens toward reconstruction.

Interlude IV

When Explanation Returns to Its Source

By the end of this part, the drift has been fully exposed—not as a failure of physics, but as a change in explanatory posture.

Nothing collapsed. Nothing contradicted experiment. Nothing demanded emergency repair.

And that is precisely why the drift endured.

Physics became extraordinarily good at *saying things that worked.* Equations spoke with authority. Language grew fluent. Metaphors settled into comfort. The discipline learned how to move forward without always looking back at the structures that first made its success possible.

This interlude exists to pause at that threshold.

What we have uncovered so far is not an error to be corrected, but a choice to be revisited. The choice to let symbols and words carry explanation on their own. The choice to accept predictive closure as conceptual completion. The choice to allow language to stand where mechanism once insisted on being traced.

That choice was understandable. It was productive. And it was incomplete.

Faraday stands at a unique position in this story.

Faraday never rejected abstraction—he simply never let it outrun observation. He trusted configuration over narration, boundaries over actors, and completion over confrontation. His intuition remained disciplined because he did not yet possess a language powerful enough to betray it.

But intuition alone could not finish the job.

Faraday saw structure with astonishing clarity, yet he lacked a framework capable of expressing that structure *universally*—across systems, scales, and technologies. His insights arrived before a language that could remain both precise and restrained at the same time. What followed filled that gap with power, but loosened discipline.

This is where the story turns constructive.

Next part does not attempt to overthrow fields, deny equations, or rename physics. It asks a simpler and more demanding question:

What if Faraday's intuition was structurally correct—but historically under-expressed?

What if electromagnetism does not require entities acting in space, but configurations creating unavoidable imbalance? What if induction is not a special law, but a universal response to delta creation? What if motors, transformers, back EMF, and eddy currents are not separate phenomena, but variations of the same equilibrium-seeking logic?

This is where Universal Energy Dynamics (UED) enters—not as a competing theory, but as a disciplining lens. A way to express what Faraday saw without inflating language into ontology. A way to unify electromagnetism without invoking actors that must "do" things.

Next part is not revisionist. It is restorative.

It returns explanation to where it began: with configuration, constraint, imbalance, and resolution.

Faraday was not wrong. He was early.

And now, with a framework capable of speaking universally without surrendering restraint, his intuition can finally be allowed to complete its arc.

The next chapters do not ask you to forget what physics has achieved. They ask you to see it again—without the metaphors that grew louder than the structure beneath them.

From here on, the task is not critique, but coherence.

Part V — Faraday Through UED

Nothing new is added here. What was always sufficient is finally allowed to speak without metaphor.

Chapter 11 — Electromagnetism Without Fields

Energy configuration, delta creation, and equilibrium resolution

Removing the Middleman

Electromagnetism is one of the most successful achievements in the history of physics. It predicts flawlessly, scales effortlessly, and powers the modern world. And yet, when asked *what actually causes electromagnetic behavior*, the answer is almost always a word rather than a mechanism.

The field.

This chapter begins by asking a deliberately uncomfortable question: *What if nothing needs to exist "in space" for electromagnetism to work at all?*

This is not a provocation for its own sake. It is a diagnostic move.

For over a century, fields have functioned as the middleman of explanation. When something happens, we say the field caused it. When influence propagates, the field carries it. When energy appears distributed, the field stores it. These statements feel explanatory—but they never tell us why a system could not remain still.

The field explains *that* something happens. It rarely explains *why it had no alternative.*

Universal Energy Dynamics (UED) begins by removing this middleman—not by denying mathematics, not by rejecting Maxwell, and not by discarding any experimental result. It simply asks whether the field is doing explanatory work, or whether it has quietly replaced something more basic.

- What if electromagnetism does not require an entity acting everywhere?

- What if it only requires arrangements that create unavoidable imbalance?

- What if motion, current, and induction are not driven—but forced to occur because equilibrium cannot complete statically?

Removing the field does not remove behavior. It removes narration.

In this chapter, electromagnetism is reconstructed without invoking anything that "acts," "pushes," or "fills space." Instead, it is described as a sequence of necessities: configuration creates delta, delta demands redistribution, and redistribution continues until equilibrium is achieved—or prevented.

Nothing mystical enters. Nothing predictive is lost. Nothing needs to be added.

What disappears is only the habit of explaining outcomes by naming an intermediary.

This opening scene marks a shift in posture. The goal is no longer to ask *what exists* between objects, but to ask *what arrangements leave systems with no choice but to respond*. In that shift, electromagnetism stops being a story about invisible actors and becomes a logic of structural inevitability.

The chapters that follow will show that once this logic is made explicit, Faraday's intuition becomes not only correct, but universal—and that electromagnetism, stripped of its metaphors, turns out to be far simpler, quieter, and more coherent than its language ever suggested.

11.1 — What UED Assumes (Briefly, Precisely)

Universal Energy Dynamics (UED) does not begin by naming entities. It begins by observing behavior.

The first assumption is minimal and non-negotiable: energy is primary, and systems evolve to reduce unresolved imbalance. Nothing else is required at the outset—no substances filling space, no mediating agents, no action-at-a-distance. What matters is whether a configuration permits equilibrium to complete.

UED therefore does not ask, *What is present here?* It asks, *What is unresolved here?*

The second assumption follows directly: delta precedes dynamics. Delta is not restricted to voltage in the electrical sense. It is any structurally enforced difference—spatial, temporal, or configurational—that prevents immediate equilibrium. Motion, current, induction, and rotation appear only when such delta cannot be resolved statically.

If no delta exists, nothing happens. If delta exists but can resolve locally, the system settles. If delta exists and cannot resolve, dynamics are inevitable.

The third assumption is restraint: no phenomenon requires an actor to "do" something. There is no need to assign causality to a field, force, or carrier. Systems respond because remaining still is not an option given their constraints. Behavior is not initiated; it is compelled by structure.

This also eliminates the need for action-at-a-distance. Influence does not jump across space. Redistribution occurs along permitted pathways defined by geometry, connectivity, and boundary conditions. What appears as propagation is simply the sequential completion of unresolved imbalance across those pathways.

Finally, UED treats equilibrium not as a final state, but as an organizing tendency. Systems are always seeking completion. Time, in this framework, is not an independent dimension driving events, but the duration over which delta persists. Where resolution is immediate, time disappears from experience. Where resolution is delayed, time emerges.

These assumptions are intentionally sparse.

They do not compete with Maxwell's equations. They do not redefine measured quantities. They do not add hidden layers.

They simply refuse to elevate descriptive tools into causes.

Under UED, electromagnetism is not something that happens *because a field exists.* It happens because specific arrangements create imbalances that cannot complete without redistribution. Everything else—current, induction, radiation, rotation—is a consequence of how that redistribution is allowed or constrained.

With these assumptions in place, the rest of the chapter does not introduce new mechanisms. It removes unnecessary narration and lets structure carry explanation.

What follows is not a reinterpretation of electromagnetism's results, but a reordering of its logic—one that aligns naturally with the experimental instincts of Faraday, while finally giving those instincts a universal language that remains disciplined enough not to turn metaphors into ontology.

11.2 — Configuration Before Interaction

Under UED, electromagnetism begins one step earlier than interaction. It begins with arrangement.

What matters first is not what objects are present, but how they are positioned, connected, and constrained. The same wire, the same conductor, the same magnet—or even the same amount of energy—can produce radically different outcomes solely by being arranged differently. This is not a secondary detail. It is the primary cause.

Configuration precedes interaction.

Traditional explanations often start by asking how one object influences another. UED inverts this habit. It asks: *What configuration makes stillness impossible?* Once that question is answered, interaction becomes inevitable rather than mysterious.

Consider a simple contrast. A conductor placed symmetrically in an environment may remain entirely inactive. The same conductor, rearranged into a loop, connected across a boundary, or moved relative to a constraint, suddenly supports current, motion, or induction. Nothing intrinsic has changed. No new "agent" has appeared. Only geometry and connectivity have been altered.

This reveals a critical point: behavior is not triggered by presence, but by incompatibility.

When components are arranged such that energy cannot distribute evenly, delta is created. That delta is not a quantity floating in space; it is a structural condition. The system is no longer allowed to rest. Response is no longer optional. What follows—current, motion, or redistribution—is simply the system doing the only thing left to it.

This is why UED treats interaction as derivative.

Interaction is not something that happens *between* objects. It is something that happens because the configuration forbids equilibrium. In this view, "interaction" is not a fundamental event; it is the visible trace of an unresolved arrangement.

This perspective also clarifies why field-based explanations often feel universal but thin. A field appears everywhere because it abstracts away configuration. Configuration, by contrast, is local, specific, and decisive. It explains not just *that* something happens, but *why it must happen here and not there*.

This emphasis aligns naturally with the experimental instinct of Faraday. Faraday did not search for hidden actors; he varied arrangements. He rotated systems, altered boundaries, changed orientations, and watched behavior reorganize accordingly. To him, geometry was not a backdrop—it was the mechanism.

UED simply formalizes this insight.

By placing configuration before interaction, explanation regains precision without adding entities. The system does not respond because it is acted upon; it responds because its arrangement leaves it no alternative. Once this is understood, electromagnetism stops looking like a dialogue between forces and starts looking like a logic of constraints resolving themselves.

From here, the next step becomes unavoidable: if configuration creates the problem, then delta—the unresolved imbalance embedded in that configuration—is the true trigger of all electromagnetic dynamics.

11.3 — Delta Creation: The True Trigger

If configuration is the cause, then delta is the trigger.

Under Universal Energy Dynamics, delta is not a specialized electrical term. It is a general condition: a structurally enforced difference that cannot resolve immediately. Delta exists whenever an arrangement prevents energy from distributing evenly across a system.

This is the moment dynamics begin.

Delta does not need to be large. It does not need to be labeled. It does not need to be measured first.

It only needs to exist without an allowed static resolution.

Traditional language often localizes delta as voltage, field strength, or potential difference. UED generalizes it. Delta can be introduced by:

- motion of conductors
- switching of connections
- rotation of assemblies
- deformation of geometry
- introduction or removal of boundaries
- asymmetry in constraints

In each case, nothing "acts." The system simply finds itself unable to complete equilibrium instantly.

That inability is delta.

This is why induction appears under such diverse conditions. Moving a magnet past a coil, rotating a conductor, changing current in a nearby loop, or deforming a conductive path all produce similar outcomes—not because they share a mystical link, but because they all create unresolved imbalance in how energy can distribute.

Delta is local and structural. It is not a property of space. It is not carried. It is not transmitted. It exists *because of arrangement* and persists only as long as the arrangement prevents completion.

Once delta appears, response is unavoidable.

The system must redistribute energy along permitted pathways. If those pathways allow quick resolution, the effect is transient. If they do not—because geometry, continuity, or boundary conditions block completion—dynamics persist. Current flows, motion continues, oscillation emerges.

This is the key unification.

Induction does not require a changing field. It requires a configuration that creates delta faster than it can be resolved.

This framing also dissolves artificial distinctions. It no longer matters whether delta is created by motion or switching, by mechanical rotation or electrical connection. The system does not care about labels. It responds only to whether equilibrium is denied.

Here, Faraday's intuition becomes precise. His insistence that change itself was causal was not poetic—it was structural. Change altered configuration, configuration created delta, and delta compelled redistribution.

UED gives that chain a universal form.

Delta is not an input. It is a condition.

And once it exists, electromagnetism is no longer a phenomenon waiting for a field to act. It is a system already in motion, simply finishing the only process it has left.

The next section follows this inevitability to its conclusion: once delta exists, energy must redistribute—and the manner of that redistribution defines everything we observe as electromagnetic behavior.

11.4 — Why Energy Must Redistribute

Once delta exists, redistribution is no longer optional.

Under UED, energy does not seek motion, propagation, or expression. It seeks completion. Redistribution is simply the process by which unresolved imbalance attempts to settle into equilibrium. What we observe as current, torque, induction, or radiation is not energy "doing something," but energy failing to finish immediately.

This distinction matters.

If equilibrium could be achieved locally and instantaneously, nothing dynamic would occur. The system would settle silently. Dynamics appear only when the configuration prevents completion, forcing energy to explore alternative pathways over time.

This is why redistribution is always constrained.

Energy does not move freely or arbitrarily. It redistributes along paths permitted by geometry, continuity, and boundary conditions. Conductors, interfaces, loops, gaps, and constraints do not guide energy like channels—they define where resolution is possible and where it is blocked.

When a pathway allows completion, redistribution ends. When all pathways block completion, redistribution persists.

This explains why motion and current often appear together. Rotation in a motor, for example, is not energy being converted into motion; it is motion emerging as the only remaining way to reduce imbalance. Similarly, current in a loop is not driven; it is the path through which delta can partially resolve.

Time enters here as a consequence, not a cause.

In UED, time is experienced as the duration of unresolved delta. Where redistribution completes quickly, time collapses into a negligible interval. Where redistribution is delayed—by geometry, inertia, or constraint—time stretches and dynamics become visible. Oscillation, rotation, and propagation are not processes occurring *in* time; they are the reason time is experienced at all.

This also clarifies why energy redistribution often appears continuous.

As long as delta is continuously recreated—by sustained asymmetry, rotation, switching, or enforced geometry—redistribution never finishes. The system lives in a state of dynamic equilibrium, constantly attempting completion and constantly being denied it.

Nothing external is required to "keep it going." Persistence is built into the configuration.

Seen this way, electromagnetism loses its mystique. It is not about invisible entities moving through space. It is about where equilibrium is blocked and how energy is allowed to respond. Redistribution is not an action; it is a necessity imposed by structure.

This is why UED treats redistribution as central. Once delta exists, everything that follows—current, force-like effects, motion, induction—is simply the system doing the only thing it can do next.

The next section applies this logic directly to induction, showing that what appears as a special electromagnetic law is actually a universal response to delta—independent of coils, fields, or historical terminology.

11.5 — Induction Without a Field

When induction is described without fields, something unexpected happens: it stops being a special effect and becomes obvious.

Under UED, induction does not require a changing field, a flux linkage, or a medium that "acts." It requires only one condition: delta must be created faster than equilibrium can complete statically. Everything else is detail.

This is why induction appears under so many guises.

Move a magnet relative to a conductor. Rotate a loop in a constrained geometry. Switch current in a nearby circuit. Deform a conductive path or change its boundary conditions.

Historically, these were treated as related but distinct mechanisms. Under UED, they collapse into one logic. Each act reconfigures the system, and that reconfiguration introduces delta that cannot be resolved without redistribution through the conductor.

Induced current is not caused. It is forced into existence.

The conductor does not "respond to a field." It becomes a permitted pathway for equilibrium-seeking redistribution. If the path is continuous and offers lower resistance than alternatives, energy flows there. If the path is broken or constrained, the effect weakens or disappears—not because the field is weaker, but because resolution is no longer possible through that route.

This also clarifies directionality.

Lenz's law, often described as a mysterious opposition, becomes structurally inevitable. Redistribution always proceeds in a way that reduces delta, not because of an opposing force, but because that is the only direction in which completion can occur. What looks like resistance is simply equilibrium asserting itself.

Importantly, this framing removes artificial distinctions. There is no fundamental difference between:

- motional induction and transformer induction
- mechanical change and electrical switching
- rotating systems and translating ones

All are cases where configuration changes deny equilibrium, and redistribution follows.

This is precisely what Faraday sensed intuitively. His insistence that "change" was the cause was not vague philosophy. It was an empirical recognition that whenever a system was rearranged in a way that prevented immediate settling, current appeared. He lacked a universal language for delta, but he never mistook induction for an action-at-a-distance.

UED completes that intuition.

Induction is not a phenomenon that happens *because electromagnetism exists*. It is what happens whenever equilibrium is structurally postponed.

Seen this way, coils and magnets are not essential ingredients. They are historical conveniences—efficient ways of creating and sustaining delta. The principle itself is broader, quieter, and far more general.

With induction stripped of its field-based narration, we can now see motors, generators, and every electromagnetic device as variations on the same theme: persistent imbalance seeking completion through the only paths allowed.

The next section follows that logic into motion itself—showing why motors do not need forces, and why rotation emerges as the most stable resolution when asymmetry cannot be removed any other way.

11.6 — Motors Without Forces

When viewed through UED, an electric motor is no longer a machine that *converts* energy using forces. It is a system that fails to reach static equilibrium—and therefore must move.

This reframing removes a long-standing confusion.

Traditional explanations rely on force narratives: currents experience forces, forces create torque, torque causes rotation. Each step sounds intuitive, yet none of them explains why the system cannot simply stop once motion begins. The force story describes motion; it does not justify its persistence.

UED starts where the force story ends.

In a motor, the defining feature is not the presence of current or magnetism. It is persistent asymmetry in configuration. The geometry of conductors, constraints of rotation, and continuity of pathways create a situation where equilibrium cannot be achieved in a static position. Every attempted settling produces a new imbalance.

Rotation is not chosen. It is imposed.

As the rotor turns, it is not being pushed forward. It is escaping configurations that are no longer allowed to resolve. Each incremental movement reduces one delta while creating another. The system is trapped in a loop of partial completion. Motion becomes the only stable response to an arrangement that denies rest.

This is why motors behave the way they do:

- Stop the asymmetry, and motion ceases immediately.
- Break the continuity, and rotation collapses.
- Alter the geometry, and direction or speed changes.

Nothing needs to "act" continuously. No force needs to be maintained. As long as the configuration recreates delta faster than equilibrium can finish, motion persists.

Torque, in this view, is not a cause. It is a measurement of how strongly equilibrium is being denied.

This also dissolves the mystery of efficiency and load. Adding mechanical resistance does not "fight the force." It introduces new pathways for energy redistribution. If those pathways allow delta to resolve more completely, motion slows. If they block resolution further, motion reorganizes or stalls. The system always does what its structure permits—nothing more.

Seen this way, motors are not special devices. They are canonical examples of dynamic equilibrium. They show what happens when geometry, continuity, and constraints are arranged so that stillness is no longer an option.

This perspective aligns cleanly with Faraday's experimental instinct. He did not search for pushing agents. He noticed that once a certain arrangement was achieved, motion simply refused to disappear. UED gives that observation its final form: motion in a motor is not produced—it is enforced by structure.

With forces removed from the center of explanation, motors become easier to understand, not harder. They are not energetic miracles. They are systems trapped in perpetual incompletion, rotating because equilibrium is always one step ahead and never allowed to arrive.

The next section extends this insight outward, showing how boundaries—not fields—determine where electromagnetic behavior completes, stabilizes, or disappears altogether.

11.7 — Boundaries Replace Fields

If configuration creates delta and redistribution resolves it, then boundaries decide where resolution is allowed to finish. Under UED, boundaries—not fields—are the decisive elements in electromagnetic behavior.

A boundary is any constraint that permits or forbids redistribution: a conductor–insulator interface, a closed loop, a surface, a gap, a contact, or a termination. These are not passive edges. They are active sites of completion. Where a boundary allows redistribution to settle, dynamics end. Where it prevents completion, dynamics persist.

This immediately clarifies phenomena that field language treats as special cases.

Shielding works not because a field is blocked, but because redistribution completes at the boundary. Once imbalance has found a permissive surface pathway, there is no residual need for interior response. The interior is quiet not because influence is excluded, but because equilibrium has already finished elsewhere.

Guidance works for the same reason. Wires, waveguides, and conductors do not carry something through space; they define the only routes along which resolution is possible. Change the boundary, and behavior reorganizes instantly. Break continuity, and dynamics collapse—not because a field weakens, but because the pathway for completion disappears.

This perspective also explains sensitivity to geometry.

Small changes at boundaries can have large effects because boundaries determine where delta can die. When completion is allowed early, effects are local and brief. When completion is delayed or denied, effects spread, persist, or transform into motion. The boundary sets the terms of engagement.

Historically, Faraday understood this intuitively. His experiments varied surfaces, enclosures, contacts, and arrangements precisely because he recognized that behavior reorganized at edges. He did not speak of boundaries as causes, but he treated them as decisive—long before language caught up.

Replacing fields with boundaries does not remove generality; it restores specificity. Fields flatten explanation by making everything "everywhere." Boundaries reintroduce local inevitability: why something happens here, not there; why it stops now, not later.

Under UED, the question shifts decisively:

- Not *what fills space?*
- But *where is equilibrium allowed to complete?*

Once that question is asked, electromagnetic behavior becomes predictable without invoking invisible intermediaries. The system responds exactly as its boundaries permit—no more, no less.

This reframing also unifies devices. Motors, generators, cages, shields, and guides differ not by field strength, but by how and where they allow redistribution to end. The boundary is the final arbiter.

With boundaries in place of fields, explanation regains discipline. We can now see electromagnetism as a choreography of completion—where structure creates delta, redistribution seeks closure, and boundaries decide the outcome.

The next section seals this reconstruction by showing why none of this contradicts existing equations—and why formalism remains perfectly intact once interpretation is realigned.

11.8 — Why This Still Matches All Equations

Removing fields from *explanation* does not remove them from *calculation.*

This distinction is critical.

Universal Energy Dynamics does not dispute Maxwell's equations, modify them, or compete with their predictive success. Every measurable quantity—current, voltage, impedance, radiation patterns, propagation speed—remains exactly as observed and exactly as computed. What changes is what we think those equations are doing.

Under UED, Maxwell's equations are treated as perfect summaries of redistribution behavior, not as statements about entities acting in space.

They work because they correctly encode how delta evolves under constraint.

When Maxwell writes relations between electric and magnetic quantities, he is not revealing agents. He is capturing consistency conditions: if imbalance exists in this form and cannot resolve locally, redistribution must appear in that form. The equations track *how equilibrium attempts propagate*, not *what propagates*.

Seen this way, several confusions dissolve at once.

- The "field" variables become state descriptors, not substances.
- Energy density becomes a local accounting of unresolved imbalance, not something stored in a medium.
- Propagation speed becomes the rate at which redistribution can sequentially complete, not the travel of a thing through emptiness.

Nothing in the mathematics requires an ontology.

This is why UED can coexist cleanly with formalism. Maxwell's equations never claimed to explain mechanism—they were interpreted that way later. They specify relationships that must hold if redistribution proceeds consistently through space and time. UED simply provides a structural narrative that makes those relationships inevitable rather than mysterious.

This also explains why field language feels indispensable in practice.

Fields are extraordinarily effective computational proxies. They allow engineers and physicists to calculate outcomes without reconstructing structure from scratch every time. UED does not ask us to abandon that efficiency. It asks us to remember that efficiency is not explanation.

Equations answer *what will happen*. UED answers *why the system had no alternative.*

Importantly, this reinterpretation also preserves unification. Electromagnetic waves remain real phenomena—not as traveling fields, but as sequential redistribution patterns sustained by persistent delta. Radiation remains predictable. Wave impedance remains exact. Nothing empirical is sacrificed.

In this sense, UED does not replace Maxwell. It rescues him from unintended metaphysics.

It allows us to say, with full confidence:

- Maxwell's equations are correct.
- Field variables are indispensable.
- Ontological inflation was optional.

This position also brings us back into alignment with Faraday. He trusted behavior without inventing actors. Maxwell formalized that behavior without demanding belief in a medium. UED completes the arc by restoring discipline between symbol and structure.

This section therefore seals the reconstruction.

Nothing new is added to physics. Nothing useful is taken away.

What changes is the direction of explanation: from entities doing things → to structures making outcomes unavoidable.

With that direction restored, electromagnetism becomes simpler, quieter, and more coherent—exactly as Faraday always suspected it was.

The final section of this chapter turns to him directly, showing why his intuition aligned so naturally with UED—and why, despite that alignment, he could not yet finish the story alone.

11.9 — Why Faraday Was Right, and Why He Could Not Finish It

Faraday was right about what mattered. He was limited only in how much language his time allowed him to build.

His experiments consistently pointed to the same truths that UED now formalizes:

- Behavior followed arrangement.
- Change, not substance, was decisive.
- Geometry and continuity governed outcomes.
- Motion and induction emerged when equilibrium was denied.

What Faraday lacked was not insight, but a generalized structural vocabulary. He could show that lines of force *behaved* as if structure mattered, but he could not yet say that structure was the cause. He could demonstrate induction everywhere change occurred, but he could not name delta as a universal condition. He could sense that fields were not things, yet had no alternative explanatory scaffold robust enough to replace them.

So he stopped where honesty demanded.

Faraday refused to pretend he knew more than he did. He offered visualizations, not ontologies. He spoke in sketches and experiments, not declarations about what "exists." In doing so, he protected physics from premature certainty—but he also left an interpretive vacuum.

That vacuum was later filled by formalism.

Mathematics stepped in to stabilize prediction, but in doing so, it quietly reified its symbols. What Faraday had treated as intuitive guides became assumed actors. Lines of force hardened into fields. Configuration receded behind abstraction. The language that was meant to assist understanding slowly replaced it.

UED does not correct Faraday. It completes his restraint.

It provides the missing bridge: a way to speak about structure, delta, redistribution, and boundaries without inventing substances or forces. It says explicitly what Faraday implied implicitly—that electromagnetism is not about things acting across space, but about systems that cannot finish equilibrium in the arrangements we impose.

Faraday could not finish this story because the scientific culture around him had no patience for mechanism without mathematics, and no vocabulary for structure without entities. Today, we do.

Seen this way, Faraday's greatest contribution was not a discovery, but a discipline:

- Do not explain before you understand.
- Do not name what you cannot justify.
- Let behavior lead language, not the other way around.

UED stands firmly in that discipline.

Faraday was right to stop where he stopped. Physics was wrong to treat that stop as an invitation to invent.

With this chapter complete, the arc becomes clear. We can now move forward—not to reject electromagnetism as it exists, but to re-anchor it in the logic Faraday always trusted. The next chapter makes that explicit, showing how all electromagnetic devices—without exception—collapse into a single structural principle once UED is applied end to end.

Closing Transition — From Interpretation to Unification

This chapter marks a decisive shift—not in physics itself, but in how physics is allowed to explain what it already knows.

By removing fields from the role of causal agents, nothing collapsed. Equations remained intact. Predictions remained exact. Devices behaved exactly as they always had. What changed was the direction of explanation: away from entities that "do things" and back toward structures that leave systems no choice but to respond.

What now stands is a cleaner logic:

- Configuration creates incompatibility.
- Incompatibility creates delta.
- Delta demands redistribution.
- Redistribution continues until equilibrium completes—or is structurally denied.

Electromagnetism, seen this way, is no longer a collection of effects tied together by historical terminology. It is a single equilibrium-seeking process expressed differently depending on geometry, continuity, and boundaries.

This is why the language of fields eventually became unnecessary. It solved communication problems—but obscured inevitability.

With UED, the explanatory load shifts back where it belongs: not on what exists in space, but on why certain arrangements cannot remain still. Once that question is asked honestly, motors, generators, induction, shielding, radiation, and confinement stop appearing as separate phenomena. They become variations of one structural truth.

This chapter has restored that truth without asking you to forget anything physics has achieved.

What follows now is the payoff.

Next chapter takes this logic and applies it universally—showing that motors, transformers, back EMF, eddy currents, and every practical electromagnetic system obey the same delta–redistribution–completion framework, without exception and without special rules.

And then, in chapter ahead, the story turns reflective once more—not to diminish Faraday, but to locate precisely where intuition alone had to stop, and where a disciplined language finally allows it to continue.

From here on, the book no longer critiques interpretation. It demonstrates coherence.

The pieces are now aligned. The vocabulary is restrained. The mechanism is explicit.

Electromagnetism is ready to be seen again—not as a theory of invisible actors, but as a logic of unavoidable response.

Chapter 12 — Reinterpreting Induction Universally

Motors, transformers, back EMF, eddy currents—one logic

One Effect, Many Names

Walk into any physics classroom, engineering lab, or textbook, and induction appears fragmented by design.

In one chapter, it is taught as motional induction—a conductor cutting magnetic flux. In another, it becomes transformer action—a changing current producing voltage elsewhere. Later, it reappears as back EMF, framed as resistance to motion. Then again as eddy currents, treated almost as a side effect or loss mechanism.

Different diagrams. Different rules of thumb. Different mental compartments.

And yet, the outcome is always the same: current appears where equilibrium cannot remain complete.

This chapter begins by challenging a quiet assumption that has shaped generations of understanding—that these are fundamentally different phenomena. They are not. They only *look* different because they were discovered through different devices, named through different experiments, and taught through different historical pathways.

Induction did not fragment in nature. It fragmented in language.

Every so-called "type" of induction begins the same way. A system that was previously able to settle is reconfigured—by motion, by switching, by rotation, by load change, or by boundary alteration.

That reconfiguration introduces delta: an imbalance that cannot resolve immediately. Once delta exists, redistribution follows. Whether we call the result torque, voltage, braking, heating, or opposition depends entirely on *where* and *how* redistribution is allowed to complete.

The devices differ. The logic does not.

This is precisely why Faraday never tried to catalogue induction into types. He sensed—without formal language—that induction was not a collection of tricks but a single behavior appearing under many constraints. Change mattered. Arrangement mattered. Boundaries mattered. Names did not.

Universal Energy Dynamics simply finishes that intuition.

This chapter does not introduce a new principle. It removes unnecessary partitions. Motors, generators, transformers, back EMF, and eddy currents will be treated not as separate inventions, but as expressions of one equilibrium logic operating under different structural conditions.

Once this is seen clearly, induction stops being something you memorize case by case. It becomes something you recognize instantly—because the system has already told you what it must do.

From here on, the question is no longer *which kind of induction is this?* It becomes: where is equilibrium being denied, and how is redistribution allowed to finish?

Everything else follows.

12.1 — The False Taxonomy of Induction

The first obstacle to understanding induction is not complexity—it is classification.

Historically, induction was divided according to *apparatus*, not *mechanism*. When current appeared in a moving wire, it was called *motional induction*. When it appeared in a secondary coil, it became *transformer induction*. When it opposed motion, it was labeled *back*

EMF. When it heated metal plates, it was dismissed as *eddy currents*. Each name reflected the experimental setup, not the underlying logic.

This taxonomy felt natural because discovery was incremental. Each phenomenon emerged in a different laboratory context, with different tools and practical concerns. Engineers needed categories to design machines. Educators needed compartments to teach. Over time, these compartments hardened into conceptual boundaries.

But none of these boundaries exist in nature.

What all these cases share is quietly ignored by the taxonomy. In every instance:

- a system that was previously settled is reconfigured,
- equilibrium is denied,
- and redistribution appears along permitted paths.

The labels obscure this unity by focusing attention on *where* the effect is seen rather than *why* it must occur. As a result, students are taught to recognize devices instead of recognizing dynamics. They learn to ask, "Is this motional or transformer induction?" instead of asking, "What structural change has prevented equilibrium here?"

This misclassification also reinforces unnecessary mystery. When different names are given to the same inevitability, the mind assumes different causes must be at work. New rules are introduced. Exceptions multiply. What should be one explanation becomes a patchwork of laws, each justified by historical precedent rather than structural necessity.

Ironically, this fragmentation runs directly against the experimental instinct of Faraday himself. Faraday did not discover "types" of induction. He discovered that whenever systems were altered in ways that denied immediate balance, electrical effects followed. The diversity of appearances never convinced him that the cause was diverse.

The false taxonomy persisted because it was useful, not because it was true.

Universal Energy Dynamics exposes this clearly. Once induction is understood as a response to delta created by configuration change, the need for multiple categories disappears. Motional, transformer, self-induction, back EMF, and eddy currents are no longer separate entries in a catalog—they are the same rule applied to different geometries and constraints.

This section clears the ground by removing that catalog.

Only then can induction be seen for what it is: not a family of effects, but one structural inevitability wearing many historical names.

12.2 — The Single Trigger: Delta Creation

Once the false taxonomy is set aside, induction reveals a striking simplicity. Across every device and every so-called "type," only one trigger ever appears: delta creation.

Delta is not a voltage, not a flux, and not a field condition. It is the structural impossibility of immediate equilibrium created by reconfiguration. Induction begins the moment a system is arranged such that energy cannot finish distributing statically.

This is why motion, switching, rotation, and load variation all work interchangeably.

- Move a conductor: geometry changes faster than equilibrium can settle.
- Switch a current: boundary conditions change abruptly.
- Rotate a system: asymmetry is sustained continuously.
- Add load: redistribution pathways are altered mid-resolution.

Different actions. Same consequence.

In each case, the system is not "acted upon." It is placed in a state where completion is denied. Delta is not injected; it is *created by constraint*. Once present, redistribution must follow, because nothing else is allowed.

This is where conventional language quietly misleads. "Changing magnetic flux" describes what accompanies delta, not what causes induction. Flux change is an accounting signal—useful for calculation, silent about mechanism. The cause lies earlier, at the moment configuration prevents balance.

Seen through UED, the trigger is brutally consistent:

Induction occurs whenever a configuration change creates delta faster than equilibrium can complete.

No exception has ever been observed.

This also explains why induction disappears in steady states. When geometry, connections, and constraints stop changing, delta stops being created. Redistribution finishes. Current fades—not because a field vanishes, but because there is nothing left to resolve.

The elegance of this trigger was intuitively clear to Faraday long before formal language existed for it. He insisted that *change itself* was causal, even when he could not say why. UED provides the missing precision: change matters only insofar as it creates unresolved imbalance.

This single-trigger view collapses complexity without loss.

- It explains why induction is indifferent to scale.
- It explains why devices differ only in geometry.
- It explains why mathematics works universally while explanation drifted.

Once delta is recognized as the trigger, induction stops being an electromagnetic curiosity and becomes a general rule of constrained systems.

The next section follows naturally: if delta is the trigger, then conductors are not recipients of induction—they are the routes through which resolution is permitted.

12.3 — Conductors as Redistribution Pathways

Once delta is created, the system faces a single question: where can equilibrium attempt to complete? The answer to that question is never "in space." It is always through structure.

Under UED, conductors do not *receive* induction. They do not respond to an external agent. They function as permitted pathways—routes along which unresolved imbalance is allowed to redistribute.

This distinction is subtle but decisive.

A conductor is not special because it carries something. It is special because it allows completion to be attempted. Where no such pathway exists, delta may still be present, but it remains trapped—producing no current, no motion, no measurable effect. Where a pathway opens, redistribution immediately appears.

This is why continuity matters more than material.

A perfectly conductive segment, if isolated or broken, does nothing. A less conductive path, if closed and continuous, becomes active.

Induction follows permission, not purity.

Loops are particularly important because they allow redistribution without termination. In a loop, energy can circulate in an attempt to settle without immediately exhausting its options. This is why coils, rings, plates, and bulk conductors repeatedly appear in induction phenomena. They are not magical geometries; they are simply structures that delay completion.

Direction, magnitude, and persistence all emerge from geometry.

- Change the loop area, and the redistribution pattern changes.
- Alter thickness or shape, and dissipation reorganizes.
- Break the loop, and induction collapses instantly.

Nothing "pushes" the current. The system is merely exploring the only remaining route that equilibrium has not yet exhausted.

This reframing also dissolves the confusion around locality.

Induction does not jump from one object to another. It unfolds across a shared structural context. When two conductors are coupled, it is not because something passes between them, but because their boundaries jointly define where redistribution can complete. Shared geometry replaces interaction-at-a-distance.

This was instinctively understood by Faraday. His experimental variations focused obsessively on continuity, contact, and enclosure. He did not ask what acted on a wire; he asked how rearranging the wire changed what was allowed to happen. He treated conductors as participants in structure, not as targets of influence.

UED simply states this explicitly.

Conductors are not sites of induction. They are the grammar of resolution.

Once this is accepted, many puzzles vanish. Back EMF, eddy currents, and coupling effects no longer require special explanations. They are simply different ways in which redistribution explores the paths we provide—and is blocked where we do not.

The next section builds directly on this insight, showing how motors and generators—often treated as opposites—are in fact the same redistribution system viewed under reversed constraints.

12.4 — Motors and Generators: Same Structure, Reversed Constraints

The traditional separation between motors and generators is one of the most persistent illusions in electromagnetism. One is said to *consume* electrical energy to produce motion; the other to *produce* electrical energy from motion. Different arrows. Different block diagrams. Different chapters.

Under UED, this distinction collapses immediately.

A motor and a generator are not opposite machines. They are the same structural system, operating under reversed boundary constraints.

What defines both is not direction of energy flow, but the same three ingredients:

- a configuration that continuously creates delta,
- pathways that permit redistribution,
- and constraints that prevent static equilibrium.

In a motor, electrical input establishes an asymmetry that cannot resolve statically. The system responds by redistributing energy through motion—rotation becomes the most effective way to reduce, but never eliminate, delta. In a generator, mechanical input establishes the asymmetry instead. The redistribution appears electrically because that is the permitted pathway.

Nothing fundamental changes. Only which boundary is forced.

This is why reversing connections turns motors into generators without altering structure. It is why regenerative braking works seamlessly. It is why the same machine can switch roles dynamically without reconfiguration. The equations tolerate this because the physics is indifferent to labels. The system simply follows the path of least incomplete resolution.

Force-based narratives struggle here. They must invent opposing causes—forces that drive, forces that resist, EMF that helps, EMF that opposes. UED requires none of this. There is only delta management. When redistribution through motion is easier, rotation dominates. When redistribution through current is easier, electrical output dominates.

This also clarifies why load matters.

Adding load does not "fight" the motor or "draw" from the generator. It alters the boundary conditions under which redistribution completes. As those conditions change, the system reallocates how delta is reduced—between motion, heat, and current—until a new dynamic equilibrium is found.

This is exactly why back EMF emerges naturally in both modes. It is not a counteracting agent. It is the system signaling that equilibrium is being approached along one pathway, forcing redistribution to slow or redirect.

Although he lacked the language to unify these roles formally, Faraday sensed this symmetry. His experiments treated motion and current as interchangeable ways of provoking the same response. The later insistence on separating motors and generators was pedagogical convenience, not physical necessity.

Under UED, the conclusion is unavoidable:

A motor is a generator whose electrical boundary is forced.

A generator is a motor whose mechanical boundary is forced.

Both are systems trapped in sustained non-equilibrium, redistributing energy until constraints change.

The next section follows this logic into one of the most misunderstood ideas in electromagnetism—transformer action—showing that even without motion, the same boundary-driven redistribution governs everything that follows.

12.5 — Transformers Without Flux Mysticism

Transformers are often presented as the most "field-dependent" of all electromagnetic devices. Energy is said to leap invisibly from one coil to another through changing magnetic flux, as if space itself were the active participant. Under UED, this picture is not wrong in calculation—but it is misleading in explanation.

A transformer works for the same reason everything else in induction works: delta is created and equilibrium is denied.

When current in the primary winding changes, the system's boundary conditions change abruptly. The shared core geometry and spatial arrangement ensure that this change cannot complete locally within the primary circuit alone. Delta appears—not in space, but in the combined configuration of primary, secondary, and core.

The secondary winding is not "influenced." It is offered as a permitted pathway.

If the secondary is open, redistribution cannot complete there. Delta remains largely unresolved, and only minimal effects appear. Close the secondary, and a new route for equilibrium becomes available. Redistribution immediately reorganizes, and current flows—not because energy jumped coils, but because completion was finally allowed elsewhere.

This also explains load dependence cleanly.

Attach a load to the secondary, and the system's completion pathways change again. Delta that previously resolved minimally must now redistribute more fully. The primary responds automatically—not in reaction, but in structural necessity. What textbooks describe as "reflected impedance" is simply the system rebalancing how and where equilibrium is attempted.

Turns ratios follow the same logic. More turns do not amplify anything mysteriously. They reshape the geometric constraints under which redistribution proceeds. Voltage and current ratios emerge because geometry dictates how imbalance is partitioned—not because flux scales energy magically.

Seen this way, nothing passes *through* the core. The core exists to delay and guide redistribution, ensuring that completion prefers the coupled structure over isolated resolution. It is a boundary-management tool, not a conduit.

This framing also resolves why transformers fail in steady DC. Without change, delta is no longer created. Redistribution completes. Silence follows. No field needs to collapse; no mechanism breaks. The system simply finishes what it was doing.

Although he did not build the transformer as we know it, Faraday already understood its essence. His ring experiments demonstrated that current in one loop reorganized behavior in another only when change persisted. He never claimed energy transfer through space—only that systems coupled by structure could not behave independently.

UED makes that insight explicit.

A transformer is not a device that transfers energy across space. It is a device that forces redistribution to complete across shared boundaries.

Once this is seen, transformers lose their aura of exception. They are simply induction without motion—proof that movement is not required, only structural denial of equilibrium.

The next section follows this inevitability into back EMF, showing that what appears as opposition is nothing more than equilibrium asserting itself at the moment it nears completion.

12.6 — Back EMF: Equilibrium Asserting Itself

Back EMF is often introduced as a kind of adversary—a counter-voltage that *fights* the source, resists motion, and limits current. The language is dramatic, but the mechanism is mundane.

Under UED, back EMF is not opposition. It is completion approaching its limit.

When a motor or generator operates, delta is continuously created by enforced asymmetry. Redistribution follows through permitted pathways—electrical and mechanical—until equilibrium is nearly achieved. At that point, something important happens: the very act of redistribution reduces the original imbalance.

That reduction is back EMF.

Nothing pushes against anything. There is no hidden reaction force. The system is simply running out of unresolved imbalance along that particular path. As redistribution becomes more effective, less delta remains to drive further flow. The measured voltage opposing the source is not a counter-agent; it is a structural signal that equilibrium is being partially satisfied.

This explains several long-standing observations cleanly.

- Increase speed in a motor, and back EMF rises—not because motion resists electricity, but because motion resolves delta more efficiently.
- Increase load, and back EMF falls—not because resistance weakens a force, but because equilibrium is now harder to approach along the same pathway.
- Remove load entirely, and current drops—not because the source is opposed, but because there is very little left to resolve.

In generators, the same logic applies. As electrical output grows, mechanical input must increase—not to overcome opposition, but to recreate delta that redistribution is successfully removing. Back EMF is simply the bookkeeping of this balance.

Seen this way, back EMF is not a special electromagnetic effect. It is the universal signature of a system nearing completion under constraint. Any process that redistributes imbalance will exhibit a self-limiting behavior once resolution becomes effective. Electromagnetism merely makes this visible and measurable.

This is why back EMF is essential for stability. Without it, redistribution would overshoot, oscillate violently, or collapse unpredictably. With it, systems settle into controlled dynamic equilibrium—smooth rotation, regulated current, predictable power delivery. Stability is not imposed from outside; it emerges from equilibrium logic.

Although the term did not exist in his time, Faraday clearly sensed this behavior. He noted that systems seemed to "answer" change proportionally, never running away unless constraints were removed. He described behavior, not battle. Later language turned that behavior into opposition.

UED restores the quieter truth.

Back EMF is not resistance to action. It is confirmation that action is working.

It marks the point where redistribution begins to succeed—and therefore must slow unless new imbalance is introduced.

The next section shows what happens when redistribution finds pathways unintentionally, revealing induction in its most unplanned form: eddy currents, where geometry alone creates opportunity and dissipation follows automatically.

12.7 — Eddy Currents: Induction Without Intention

Eddy currents are often treated as a nuisance—losses to be minimized, side effects to be suppressed, heat to be managed. Under UED, they reveal something far more fundamental: induction does not require design, purpose, or even awareness of circuitry. It requires only opportunity.

Whenever delta is created in the presence of a conductive bulk, redistribution will occur along any path that geometry permits. If that path happens to close on itself—even accidentally—current flows. That is all an eddy current is: unplanned completion.

There is no coil. No terminals. No intention.

A moving magnetic assembly near a metal plate, a rotating conductor in an asymmetric environment, or a changing current adjacent to a slab of metal all create delta. The bulk material offers countless microscopic loops—closed pathways formed by continuity alone. Redistribution explores them immediately.

Heat, braking, and damping follow not because energy is being wasted, but because equilibrium is being reached locally and irreversibly. Where a designed loop delays completion, eddy currents accelerate it. The result is dissipation.

This explains their defining features:

- **Localization**: Eddy currents stay near surfaces and regions where geometry allows closure.
- **Speed dependence**: Faster change creates delta faster than local completion, intensifying currents.
- **Material dependence**: Better conductors allow easier completion, increasing effect.
- **Geometry sensitivity**: Slots, laminations, and segmentation disrupt loops and suppress currents—not by blocking fields, but by breaking completion pathways.

Nothing mystical is happening. The system is simply finding too many easy ways to finish.

This is why eddy currents are so useful as brakes. Motion itself creates delta, and the metal offers abundant routes for redistribution. The resulting currents dissipate energy as heat, rapidly reducing imbalance. Motion slows—not because it is opposed by a force, but because equilibrium is being achieved efficiently.

Seen this way, eddy currents are not a special phenomenon at all. They are induction stripped of ceremony. No coils. No terminals. Just structure doing what structure always does when imbalance appears.

This perspective aligns perfectly with the experimental instinct of Faraday. He treated conductive masses as active participants, not passive victims of influence. He varied shape, thickness, and continuity, and watched behavior reorganize accordingly. He understood—long before the language existed—that geometry decides fate.

UED makes this explicit.

Eddy currents demonstrate the universality of induction more clearly than any textbook device. They prove that:

- induction does not belong to circuits,
- it does not require design,
- and it does not wait for permission.

Wherever delta appears and completion is allowed, redistribution follows.

The next section completes this universal picture by addressing time and frequency—showing that induction persists not because of oscillation itself, but because equilibrium is being denied repeatedly in time.

12.8 — Time, Frequency, and Persistence

One of the quiet confusions surrounding induction is the role of time. Frequency is often treated as a driving parameter, oscillation as a requirement, and AC as a fundamentally different regime. Under UED, time enters the story much later—and far more simply.

Time appears only because equilibrium is not allowed to complete.

Induction is not sustained by oscillation. Oscillation is sustained because delta is recreated repeatedly. Frequency does not cause induction; it counts how often equilibrium is denied.

This distinction clarifies several puzzles at once.

A steady DC system shows no induction not because DC is special, but because once configuration settles, delta stops being created. Redistribution finishes. Silence follows. Introduce switching, rotation, or modulation, and delta reappears immediately—even in a DC system. Induction resumes, not because AC has arrived, but because completion has been interrupted.

In alternating systems, each cycle recreates imbalance before the previous redistribution can finish. The system never settles. Induction persists as long as reconfiguration outruns completion. Frequency simply sets the rate of denial.

Higher frequency does not "strengthen" induction by magic. It reduces the time available for redistribution to complete locally. As a result, imbalance spreads more widely, currents reorganize, and new pathways dominate. Skin effect, proximity effect, and frequency-dependent losses all follow naturally from this logic. They are not wave phenomena in the ontological sense; they are consequences of rushed completion.

Time, in this framework, is not a backdrop. It is a measure of unfinished business.

Where redistribution finishes quickly, time collapses into insignificance. Where it is repeatedly interrupted, time stretches into cycles, oscillations, and persistence. Induction exists *because* time exists in that system—not the other way around.

This is why induction scales seamlessly from slow mechanical motion to radio-frequency behavior. The system does not care about speed in the abstract. It cares only about whether delta is being recreated faster than it can be resolved.

Once again, this was sensed intuitively by Faraday. He observed that steady arrangements went quiet, while changing ones came alive—regardless of whether the change was slow or fast. He did not attribute this to oscillation; he attributed it to ongoing change.

UED sharpens that intuition.

Induction persists not because something vibrates, but because equilibrium is repeatedly postponed.

The final section of this chapter gathers everything together, showing why all induction phenomena—across devices, scales, and regimes—must behave identically once this logic is accepted.

12.9 — Why All These Behave Identically

At this point, the unification is no longer philosophical—it is unavoidable.

Across motors, generators, transformers, back EMF, eddy currents, and frequency-dependent effects, the same structural sequence appears every time:

1. Configuration changes
2. Delta is created (equilibrium cannot complete immediately)
3. Redistribution seeks resolution along permitted pathways
4. Boundaries determine where and how completion occurs

Every so-called "different" induction phenomenon is simply a different way of maintaining or interrupting this sequence.

What varies from device to device is not the mechanism, but:

- how delta is created (motion, switching, rotation, loading),
- how long it is sustained,
- and where completion is allowed to finish.

A motor maintains delta through enforced geometric asymmetry. A generator maintains it through mechanical constraint. A transformer maintains it through boundary-coupled reconfiguration. Back EMF appears when redistribution becomes effective. Eddy currents appear when completion is too easy and uncontrolled.

Different outcomes, identical logic.

This is why no special rules are ever required. The same equations apply everywhere, not because electromagnetism is versatile, but because equilibrium logic is universal. The mathematics works precisely because it tracks redistribution consistently, even when interpretation wandered.

Once this is seen, several long-standing confusions dissolve immediately:

- There is no need for multiple kinds of induction.
- There is no hierarchy of "primary" and "secondary" effects.
- There is no contradiction between electrical, mechanical, and thermal outcomes.

They are all expressions of the same unfinished equilibrium, resolving through different structural exits.

This also explains why induction appears far beyond classical machines—at microscopic scales, in bulk materials, in fast switching electronics, and in electromagnetic radiation. The mechanism does not belong to coils or magnets. It belongs to any system where structure denies immediate balance.

This is precisely why Faraday trusted intuition over taxonomy. He did not need unification to be declared; he observed that nature behaved the same way regardless of apparatus. What he lacked was not insight, but a framework willing to say, plainly and universally:

All induction is the same phenomenon, because all equilibrium denial is the same problem.

Universal Energy Dynamics supplies that framework.

With this section, induction is no longer a chapter in electromagnetism. It is a general law of constrained systems, visible most clearly in electrical phenomena but not limited to them.

The closing transition now follows naturally: having shown that one logic governs everything, the book can finally step back and ask the reflective question—not *how induction works*, but why it took so long to see that it always worked the same way.

Closing Transition — One Logic, Everywhere

This chapter completes the unification that induction has quietly demanded since its discovery.

What once appeared as a scattered collection of effects—motional induction, transformer action, back EMF, eddy currents, frequency dependence—now stands revealed as one unavoidable sequence. Nothing new was added. Nothing empirical was revised. The only change was the willingness to ask a more honest question: *Where is equilibrium being denied, and how is resolution permitted?*

Once that question is asked, induction stops behaving like a set of rules tied to specific machines. It becomes a general response of structured systems—a rule that appears wherever configuration change outpaces completion.

This is why the same logic governs:

- a slowly rotating motor,
- a high-frequency transformer,
- a braking metal plate,
- and a fast-switching electronic circuit.

They differ in scale and intent, not in mechanism.

Seen this way, induction is no longer a special chapter of electromagnetism. It is a case study in equilibrium logic, made visible because electrical systems expose redistribution so cleanly. What we learned here applies wherever structure, constraint, and change intersect.

This also reframes Faraday's legacy with clarity and fairness. Faraday was not missing a piece of electromagnetism. He was missing a language that science itself had not yet matured enough to offer. His insistence that change mattered was correct. His refusal to invent invisible agents was disciplined. What remained unfinished was not his insight, but the framework needed to state it universally.

Universal Energy Dynamics provides that framework—not as a replacement for physics, but as a completion of its restraint.

With induction now unified, the final chapter can turn reflective rather than technical. Next chapter steps back to ask what Faraday revealed, what he could not yet articulate, and why modern physics, despite immense mathematical success, drifted away from the simplicity he trusted.

The story now moves from mechanism to meaning.

Not to add another theory— but to understand why clarity so often arrives before language is ready to receive it.

Chapter 13 — Why Faraday Was Right but Incomplete

What intuition revealed and what language could not yet express

The Silence After Discovery

There is a peculiar stillness that follows genuine discovery.

By the time Faraday reached the later phase of his scientific life, the foundational pillars of electromagnetism were already in place. Motion from electricity. Induction from change. Coupling without contact. Boundaries that stabilized interiors. The laboratory had spoken clearly.

And yet, Faraday did not rush to explain what he had uncovered.

He wrote less. He speculated cautiously. He resisted the growing pressure to declare what electricity *was*, what magnetism *consisted of*, or what invisible medium must exist to make the equations behave. In an era increasingly confident in abstraction, his restraint appeared almost anachronistic.

This silence has often been misread.

Some interpret it as limitation—an experimentalist unable to climb the final theoretical step. Others frame it as deference, suggesting that Faraday simply left the "real thinking" to mathematicians. Both readings miss the deeper truth. Faraday was not silent because he lacked ideas. He was silent because he refused to name what he could not justify structurally.

His discoveries had reached a boundary—not of evidence, but of language.

Faraday knew what change did. He knew what arrangement allowed. He knew how systems behaved when equilibrium was denied. What he did not have was a vocabulary that could express these facts without inventing entities, substances, or metaphysical carriers. Rather than cross that line, he stopped.

This pause was not hesitation. It was discipline.

Faraday sensed—correctly—that premature explanation would do more harm than honest incompletion. He chose fidelity to observation over the comfort of closure. In doing so, he left behind a body of work that was experimentally complete but conceptually open, waiting for a future framework capable of speaking without fabrication.

This chapter begins in that quiet space.

Not to criticize what Faraday did not say, but to understand why he was right to stop, and why completion required a language that science itself had not yet learned how to speak.

13.1 — What Faraday Saw Clearly

Faraday's clarity did not come from theory. It came from repeated confrontation with behavior.

Across decades of experimentation, a consistent pattern emerged in his work—one that he never reduced to formula but never lost sight of either. He saw that electrical and magnetic phenomena did not originate from substances, forces, or actions at a distance. They arose from how systems were arranged and how that arrangement was changed.

The first insight he grasped with remarkable firmness was that change itself was causal. Static situations, no matter how intense, produced no new effects. Motion, switching, rotation, and reconfiguration did. This was not a philosophical claim for Faraday; it was an experimental fact. Whenever something happened, something had changed first.

Closely tied to this was his recognition that geometry mattered. Orientation, continuity, enclosure, and proximity consistently determined outcomes. Rotate a wire, and behavior reorganized. Break a loop, and effects vanished. Enclose a region, and its interior stabilized. These were not secondary details; they were decisive. Faraday treated spatial arrangement as an active participant, long before physics had language for configuration-driven causality.

He also saw, perhaps more clearly than anyone before him, that boundaries controlled behavior. His work on shielding and induction revealed that where a system was allowed to "finish" mattered more than what acted upon it. Effects appeared or disappeared at surfaces, contacts, and enclosures. Faraday followed edges and interfaces with the same seriousness others reserved for forces.

Most importantly, he understood that motion, induction, and stability were not separate phenomena. They were expressions of the same underlying condition. Systems that could not settle responded with motion or current. Systems that were allowed to settle went quiet. There was no mystery in this for him—only consistency.

What Faraday saw, in essence, was structure before substance.

He did not articulate it that way, because the language did not yet exist. But his experiments consistently privileged arrangement over agency, change over presence, and completion over action. These were not accidental insights; they were the result of a mind trained to trust what repeated observation made unavoidable.

Faraday knew something fundamental was at work. He simply refused to pretend he knew more than he did.

The next section explores the other side of this clarity—not what Faraday understood, but what he deliberately refused to invent, even when the scientific culture around him was eager for answers at any cost.

13.2 — What He Refused to Invent

Faraday's greatest intellectual strength was not only what he perceived, but what he declined to assert.

At a time when scientific explanation was drifting toward invisible substances and metaphysical carriers, Faraday resisted the temptation to fill gaps with names. He encountered electrical and magnetic effects everywhere, yet he refused to declare what electricity *was*, what magnetism *consisted of*, or what unseen medium must exist to make influence possible.

This refusal was not hesitation. It was method.

He rejected the language of electric fluids, magnetic effluvia, and hidden forces—not because they were provably false, but because they were structurally unjustified. They described outcomes without clarifying necessity. They soothed curiosity without deepening understanding. Faraday sensed that once such language hardened, it would become an obstacle rather than a bridge.

Even his famous "lines of force" were never offered as entities. They were visual aids—ways of seeing structure and relation. He drew them, manipulated them, and reasoned with them, but he never claimed they were things inhabiting space. The moment a metaphor risks being mistaken for a mechanism, Faraday stepped back.

He also resisted action-at-a-distance explanations. Effects that appeared separated in space did not persuade him that something traveled invisibly between them. Instead, he focused on what the system as a whole permitted. If behavior reorganized, he assumed configuration and continuity were responsible—not a messenger acting across emptiness.

This discipline isolated him.

While others grew increasingly confident in abstraction, Faraday remained grounded in experiment. He was willing to let explanation remain incomplete rather than become careless. In doing so, he preserved a rare form of scientific integrity: the courage to say *"I do not yet know how to speak about this without distortion."*

This restraint is often misunderstood as theoretical weakness. In truth, it was epistemic honesty. Faraday understood that explanation without mechanism is not progress—it is decoration.

Universal Energy Dynamics makes clear why this restraint mattered. Faraday lacked not insight, but a general language of structure—a way to speak about configuration, imbalance, and completion without inventing actors. Until such a language existed, silence was the only rigorous option.

This is why Faraday stopped where he did. He refused to cross the line where naming would have replaced understanding. That refusal kept physics clean, even as it left questions open.

The next section addresses the cost of this restraint: why intuition alone, however precise, could not unify electromagnetism without a vocabulary capable of generalization—and why completion had to wait for a framework Faraday himself could not yet possess.

13.3 — The Limits of Intuition Without Language

Intuition can reveal truth, but it cannot stabilize it.

Faraday's insights were exact in behavior yet fragile in transmission. He could demonstrate again and again that change produced effects, that arrangement governed outcome, that boundaries decided stability. But without a shared language to generalize these observations, each insight remained local—tied to a specific experiment, a particular setup, a familiar geometry.

This is where intuition met its natural limit.

To unify phenomena, science requires more than seeing; it requires naming relations in a way that survives translation—across devices, scales, and generations. Faraday's intuition was relational, dynamic, and geometric. The language available to him was static, substance-oriented, and force-centric. There was no vocabulary for equilibrium denial, no concept of delta as a structural condition, no grammar for redistribution as necessity rather than action.

As a result, Faraday could show *that* phenomena were connected, but not *why* they had to be.

Consider induction. Faraday knew motion, switching, and rotation all produced current. He knew the effect vanished when change ceased. Yet to generalize this beyond apparatus would have required a concept like "unresolved imbalance created by configuration." That idea did not exist. To invent it prematurely would have meant guessing at ontology—precisely what he refused to do.

This was not a personal limitation; it was a historical one.

Intuition excels at detecting patterns. Language is required to lock those patterns into necessity. Without language, insight remains vulnerable to reinterpretation. Others can describe the same behavior with different metaphors, different symbols, different assumptions—and over time, the original clarity dissolves.

This is exactly what happened after Faraday.

His demonstrations were absorbed. His restraint was not. In the absence of a structural vocabulary, later thinkers filled the gap with what language they had: fields, forces, fluxes—terms that enabled calculation but quietly reintroduced ontology. Intuition gave way to symbol, and symbol began to masquerade as mechanism.

Faraday could not prevent this drift because he had no alternative language to offer.

Universal Energy Dynamics reveals why completion had to wait. The missing piece was not more experiments or better mathematics, but a conceptual layer capable of saying: configuration creates incompatibility; incompatibility creates delta; delta compels redistribution until completion is allowed.

Without that layer, intuition could guide discovery but not unify it.

Faraday reached the edge of what could be known without inventing. He stopped there deliberately. The incompleteness was not his failure; it was the cost of intellectual honesty in an era whose language was not yet ready.

The next section shows why mathematics—despite its power—could not rescue this situation either, and why formalism alone would stabilize prediction while leaving explanation unresolved.

13.4 — Why Mathematics Could Not Save Him

It is tempting to assume that mathematics could have completed what intuition began. History often frames the story that way: Faraday discovers, Maxwell explains. But this narrative oversimplifies a deeper constraint. Mathematics did not *fail* Faraday. It was simply not designed to do the kind of work his insight required.

Mathematics excels when entities and variables are already defined. It formalizes relationships between known quantities. What Faraday had uncovered, however, was not a new quantity—it was a new way in which systems behave. His observations were about arrangement, denial of equilibrium, and structural necessity. These are relational conditions, not immediately expressible as variables.

To mathematize something, one must first decide *what exists.*

That decision was precisely what Faraday refused to make prematurely. He would not assert a medium, a fluid, a force, or a field as a physical actor. Without such commitments, mathematics had nothing stable to anchor itself to. Equations cannot operate on restraint; they require assumptions.

This is why Maxwell could proceed where Faraday stopped. Maxwell introduced abstract quantities—fields, potentials, densities—not as claims about substance, but as mathematical conveniences. That move allowed formalism to flourish. Prediction stabilized. Engineering advanced. But the cost was subtle: symbols began to stand in for mechanism.

Mathematics did its job perfectly. It encoded behavior with extraordinary precision. What it could not do—and never claims to do—is explain *why the structure leaves no alternative.* Formalism tracks outcomes; it does not justify inevitability unless the ontology is already settled.

Faraday sensed this intuitively. He understood that once symbols are introduced, they acquire authority. They shape what questions are asked and which answers are considered legitimate. To introduce mathematics without conceptual clarity would have meant locking uncertainty into equations, giving it the appearance of understanding.

So he declined.

This was not anti-mathematical. It was pre-mathematical honesty. Faraday recognized that without a language of structure—without concepts like delta, redistribution, and completion—equations would stabilize results while quietly redirecting interpretation. That is exactly what later happened.

Universal Energy Dynamics clarifies this boundary. It shows that Faraday did not need more mathematics. He needed a structural ontology that was neither substance-based nor force-based. Only once such an ontology exists can mathematics regain its proper role: not as a generator of explanation, but as a faithful recorder of necessity.

Faraday could not cross that bridge because the bridge had not yet been built.

The next section shows how, once mathematics took the lead without that bridge, physics advanced operationally—but drifted conceptually—replacing mechanism with language and inevitability with metaphor.

13.5 — How Completion Drifted After Him

Once Faraday stepped back, physics did not pause. It accelerated.

The vacuum he left was not one of data or capability, but of explanation. Experiments worked. Devices multiplied. Predictions improved. What was missing was not success, but mechanism expressed without invention. Into that gap, language quietly moved.

Fields, forces, flux, potentials—these were introduced first as computational scaffolding. They allowed mathematics to operate smoothly without committing to a specific substance. In their original intent, they were neutral tools. But over time, something subtle happened: the tools began to stand in for understanding.

Metaphor hardened into ontology.

What had once been a way of speaking about relationships became a way of asserting what exists. Fields were no longer descriptors; they became actors. Flux was no longer an accounting device; it became causal. The success of equations conferred legitimacy on the language surrounding them, and legitimacy slowly replaced scrutiny.

This was not a mistake. It was a natural drift.

Operational success reduces pressure to reinterpret. When machines work, when predictions hold, when technology advances, explanation becomes optional. The question shifts from *why must this happen* to *how reliably does it happen*. Physics, understandably, chose reliability.

But something was lost in that choice.

Faraday's discipline—his refusal to name without necessity—was not inherited along with his experiments. The next generation accepted his results but relaxed his restraint. Where he had stopped out of honesty, they proceeded out of utility. The result was a conceptual architecture that worked brilliantly while remaining structurally vague.

This is why so many electromagnetic explanations still feel circular. Fields cause forces; forces define fields. Flux changes induce EMF; EMF is defined by flux change. The language is internally consistent but externally thin. It describes behavior without revealing inevitability.

Universal Energy Dynamics makes clear that this drift was not avoidable in Faraday's time. Without a language of structure—without concepts like delta and equilibrium denial—something had to fill the explanatory space. Mathematics filled it, and metaphor followed.

Completion drifted not because Faraday was wrong, but because he was too careful to guess.

The irony is instructive. Physics did not advance *despite* Faraday's restraint; it advanced because of it. His silence preserved clarity long enough for formalism to develop. But formalism, left alone, could not complete the story. It stabilized prediction while leaving mechanism implicit.

This is why reinterpretation is now necessary—not to undo progress, but to re-anchor it.

The next section explains why this completion could not have happened earlier, and why a framework like Universal Energy Dynamics requires a level of conceptual maturity that only became possible after physics had already succeeded operationally.

13.6 — Why UED Could Not Have Appeared Earlier

It is tempting to ask why a framework like Universal Energy Dynamics did not emerge in Faraday's time—after all, the experiments were already decisive. The answer is not oversight or hesitation. It is conceptual readiness.

UED requires a way of thinking that science had not yet earned.

First, it requires treating equilibrium as primary, not motion. Classical physics was built around action—forces acting on bodies, causes producing effects. Even when Faraday observed behavior emerging from arrangement, the dominant intellectual habit still searched for agents. A framework centered on *denial of equilibrium* rather than *application of force* would have sounded inverted, even incoherent, to that era.

Second, UED assumes energy as relational, not as a substance or a transferable commodity. In Faraday's time, energy was still entangled with mechanical intuition—something possessed, stored, moved, or spent. The idea that energy dynamics are defined by configuration and constraint, rather than by ownership or flow, required decades of thermodynamic and statistical maturity that simply did not yet exist.

Third, UED treats time as a consequence, not a background. Faraday could observe persistence and decay, but the conceptual leap—that time itself reflects unresolved redistribution—belongs to a later intellectual landscape. Until entropy, irreversibility, and non-equilibrium thinking were absorbed deeply into physics, time could not be reframed this way.

Finally, UED demands discipline against ontology inflation. It insists that explanation should not outrun necessity. This stance becomes possible only after a field has already succeeded operationally. Before that success, restraint looks like obstruction. After success, restraint becomes clarity.

Faraday lived before that turning point.

His world needed invention to move forward—new terms, new symbols, new mathematical machinery. Ours is different. We inherit a physics that works extraordinarily well, yet carries explanatory discomfort it no longer needs to tolerate. Only from that vantage point can reinterpretation occur without jeopardizing utility.

This is why UED is not a correction to Faraday, nor a replacement for Maxwell. It is a late-stage clarification—something that can only appear after prediction is secure, technology is mature, and the cost of conceptual drift has become visible.

In this sense, UED could not have appeared earlier because it would have been premature. It depends on the very formal success that once obscured the need for it.

Faraday reached the edge of honest explanation and stopped. Physics moved on, built the machinery, and proved its power. Only now—standing on that accumulated success—can we return to his restraint and finish what he rightly left incomplete.

The next section shows why this return is not revisionist, but deeply faithful—revealing Faraday as the first thinker whose instincts already aligned with UED, even though the language to express it did not yet exist.

13.7 — Faraday as the First UED Thinker (Unknowingly)

Seen through the lens of Universal Energy Dynamics, Faraday does not appear as a precursor waiting to be corrected. He appears as something rarer: a thinker who was already operating within the logic of UED, without possessing its language.

Faraday never began with forces. He began with arrangement.

His experimental instinct was consistently configuration-first. When something unexpected occurred, his response was not to ask *what acted*, but to ask *what was different*. He rotated systems, altered continuity, changed boundaries, and watched behavior reorganize. This is precisely the UED move: identify how equilibrium is being denied before asking how response unfolds.

Faraday also treated change as causal, not as a modifier of something else. He did not believe motion "activated" a field or that switching "released" an influence. Change mattered because it altered what completion was possible. This aligns exactly with the

UED notion of delta creation—even though Faraday had no such term.

Most strikingly, he privileged boundaries over space. His work on shielding, enclosures, and coupling showed that behavior finished where it was allowed to finish. Interiors became quiet not because influence was blocked, but because resolution occurred elsewhere. This is boundary-governed redistribution in its purest experimental form.

Faraday's famous discomfort with abstraction now looks principled rather than naive. He resisted treating lines of force as substances because he sensed they were descriptive of structure, not carriers of action. UED formalizes this intuition by stripping explanation of entities altogether and restoring inevitability to configuration.

Even his silence fits the pattern.

Faraday stopped not because he lacked insight, but because he refused to cross into ontological guesswork. UED respects that boundary. It does not retroactively assign him beliefs about fields, media, or hidden agents. It simply gives language to what his experiments already enforced: systems behave as they must when equilibrium is structurally denied.

In this sense, Faraday was not incomplete because he was wrong. He was incomplete because honesty outpaced vocabulary.

Universal Energy Dynamics does not reinterpret Faraday against his will. It reads him faithfully—by honoring what he emphasized, what he avoided, and where he chose to stop. It reveals that the core of UED was already present in his method: configuration-first reasoning, change as trigger, redistribution as necessity, and boundaries as final arbiters.

Faraday did not articulate UED. He lived it experimentally.

The next section completes this alignment carefully, showing how UED finishes Faraday's work without correcting it—by providing language where silence was once the only ethical option.

13.8 — Completion Without Correction

Universal Energy Dynamics does not arrive to amend Faraday. It arrives to respect the boundary he refused to cross.

This distinction is essential. Completion, in this context, does not mean fixing an error or replacing an outdated view. It means providing language for truths that were already present but ethically unspoken. Faraday did not leave gaps because he misunderstood electromagnetism; he left gaps because filling them would have required inventing what he could not justify.

UED finishes the story without violating that restraint.

Where Faraday showed that change mattered, UED names delta—not as a new entity, but as a structural condition. Where Faraday demonstrated that arrangement governed behavior, UED formalizes configuration as cause. Where Faraday relied on boundaries experimentally, UED identifies completion at boundaries as the decisive mechanism.

Nothing Faraday said is contradicted. Nothing he refused is forced retroactively into his mouth.

This is why reinterpretation is not revisionism. It does not claim that Faraday "meant" UED in some hidden way. It claims something more modest and more honest: his discipline preserved clarity long enough for completion to become possible later.

Importantly, UED also avoids a common historical mistake—treating silence as ignorance. Faraday's refusal to speculate was not a lack of imagination; it was a recognition that explanation must not outrun necessity. UED aligns with that principle by refusing to add fields, forces, or carriers where structural inevitability already suffices.

In this sense, UED does not extend Faraday's intuition by adding content. It extends it by adding restraint to language—ensuring that explanation follows structure rather than metaphor.

This is why the completion feels quiet rather than dramatic. There is no overthrow of equations, no denunciation of formalisms, no new ontology to defend. The framework simply says what Faraday allowed himself to show but not to name:

Systems behave the way they do because equilibrium is being denied by configuration, and response continues until completion is allowed.

By stopping where he did, Faraday ensured that when language finally caught up, it would not have to unlearn invention first. UED honors that gift by completing the explanation without betraying the silence that protected it.

The final transition now follows naturally. Having seen why Faraday was right, and why incompletion was necessary, the book can close by stepping back—reflecting not just on electromagnetism, but on how science itself advances: insight first, language later, and restraint always as the quiet guardian of truth.

Closing Transition — When Insight Waits for Language

Faraday's story does not end in resolution. It ends in permission.

He permitted physics to move forward without forcing it to explain what it could not yet speak about honestly. He showed what behavior demanded, then stepped aside when language threatened to outrun structure. In doing so, he left behind not an unfinished theory, but a protected insight—one that could survive success without being distorted by premature certainty.

This chapter has traced that arc carefully.

Faraday saw clearly what mattered: change, configuration, boundaries, and completion. He refused to invent what he could not justify. Mathematics later stabilized prediction but quietly replaced mechanism with metaphor. Operational success followed, while conceptual clarity waited. Universal Energy Dynamics arrives not as a correction, but as the moment when language finally becomes restrained enough to match insight.

This pattern is not unique to electromagnetism. It is how science advances at its most honest:

- observation precedes explanation,
- intuition precedes formalism,
- and completion waits until invention is no longer required.

Faraday stood at the edge of that sequence and chose integrity over closure. That choice made it possible for reinterpretation to occur without undoing progress.

With this understanding, electromagnetism no longer feels like a patchwork of historical accidents. It reads as a coherent story delayed by the maturity of its language. The phenomena were always unified. What was missing was the courage to explain them without adding things.

This book now closes that gap.

Not by declaring a new theory, but by restoring a direction of explanation that Faraday trusted instinctively: from structure to necessity, from arrangement to response, from equilibrium denied to redistribution compelled.

Some truths are discovered early. Some are spoken late.

And the distance between them is not time, but discipline.

With Faraday's arc complete, the book can now step back—having shown not only how electromagnetism works under Universal Energy Dynamics, but why understanding itself often waits for language to learn restraint.

Interlude V

After Completion, Before Legacy

There is a particular quiet that follows genuine completion—not the quiet of exhaustion, but the stillness that comes when a structure finally holds.

This part did not add new phenomena to electromagnetism. It removed unnecessary weight. Fields were released from ontology. Forces were relieved of causation. Induction was returned to inevitability. What remained was not emptiness, but coherence.

At this point in the journey, something important has happened.

Faraday no longer appears as a historical figure surrounded by later interpretation. He stands revealed as a methodological anchor—a reminder that physics, at its core, is not about naming what exists, but about understanding why systems cannot behave otherwise. Universal Energy Dynamics did not reinterpret his results; it re-aligned explanation with the discipline he practiced instinctively.

This interlude exists to pause before moving forward—not because the argument needs rest, but because the direction of inquiry is about to change.

Up to now, the work has been corrective and reconstructive. We examined where interpretation drifted, how language replaced mechanism, and why unification required restraint rather than invention. From here onward, the question is no longer *what Faraday meant* or *how electromagnetism works*.

The question becomes: what does this way of understanding demand next?

If electromagnetism can be clarified by restoring intuition and structure, what does that say about modern physics more broadly? If engineering succeeded spectacularly while ontology quietly stagnated, what does that reveal about our priorities? And if Faraday's discipline turns out to be more relevant now than ever, how might he respond to today's abstractions, equations, and quantum metaphors?

This is where legacy begins—not as reverence, but as responsibility.

The final part of this book does not look backward. It looks sideways and forward. It asks how Faraday's way of thinking would confront contemporary physics, how invention overtook understanding, and why the next evolution of science may depend less on new mathematics and more on recovering intellectual restraint.

In that sense, final part is not a conclusion. It is a continuation of method.

We now step beyond reinterpretation and into relevance—carrying Faraday's discipline into the present, not to judge modern physics, but to ask whether clarity has once again outrun the language we are using to describe it.

The next chapter begins with a simple but unsettling thought experiment:

If Michael Faraday were alive today—what would he recognize immediately, and what would trouble him deeply?

That question opens the final movement of the book.

Part VI — Legacy, Relevance, And Continuity

The past does not instruct us by being correct. It instructs us by showing where restraint once protected truth.

Chapter 14 — If Faraday Lived Today

How he might view modern physics, quantum language, and abstraction

Meeting the Present Without Awe

If Faraday were alive today, he would not be dazzled by our instruments.

Particle accelerators, quantum computers, space telescopes, nanofabrication labs—these would impress him as achievements of craftsmanship, not as proofs of understanding. Faraday never confused reach with clarity. He admired precision, but he trusted behavior.

What would interest him immediately is not what we can build, but how confidently we speak about what we cannot see.

He would notice, within minutes, that modern physics is extraordinarily capable and curiously uneasy. That we predict with astonishing accuracy, yet explain with metaphors that sound increasingly detached from necessity. That our equations work perfectly while our stories about what they "mean" grow ever more elaborate.

Faraday would not reject modern physics. But he would ask it uncomfortable questions.

This chapter explores what those questions might be—and what his way of thinking reveals about where physics stands today.

14.1 — What Faraday Would Instantly Recognize

Faraday would feel at home in the laboratory.

He would recognize the discipline of experimental control, the obsession with reproducibility, the care taken in isolating variables. He would admire the way modern engineering has learned to respect constraints, design around boundaries, and exploit configuration with astonishing sophistication.

In electromagnetism, he would immediately recognize continuity.

Motors, generators, transformers, shielding, signal propagation—none of this would surprise him conceptually. The scale and refinement would astonish him, but the logic would feel familiar. Change still matters. Geometry still governs. Boundaries still decide outcomes. The machines would look different; the behavior would not.

Even in quantum experiments, Faraday would recognize something essential: patterned response under constraint.

He would not be shocked that systems behave probabilistically under certain configurations. He never expected nature to behave simply; he expected it to behave *consistently*. Repeated outcomes under controlled conditions would reassure him that physics had not abandoned discipline.

What would concern him is not the mathematics. Faraday respected mathematics deeply, even though he did not practice it fluently. He understood its stabilizing power. He would see immediately that modern formalism is elegant, powerful, and internally consistent.

What would trouble him is how quickly symbols are treated as reality.

14.2 — Where Abstraction Would Make Him Uneasy

Faraday was allergic to explanation that outran justification.

Modern physics, especially at fundamental scales, would strike him as intellectually confident but ontologically careless. He would notice how readily we speak of entities—fields, particles, wavefunctions, virtual excitations—as if naming resolved uncertainty.

He would ask questions that sound naïve but cut deep:

- *Where, exactly, is this object you are describing?*
- *What experiment forces you to believe it exists, rather than behaves as if it exists?*
- *What changes if I refuse this interpretation but keep the equations?*

Quantum language, in particular, would make him pause.

Superposition, collapse, entanglement—Faraday would not deny the phenomena. He would deny the stories we tell about them. He would sense immediately that much of modern interpretation resembles the very "electric fluids" and "magnetic effluvia" he refused to invent.

He would see that we often mistake computational success for ontological certainty.

To Faraday, a wavefunction would look like a line of force: a powerful visualization tool that becomes dangerous the moment it is treated as a thing. He would not ask whether the wavefunction is real. He would ask whether calling it real adds necessity or merely comfort.

And if the answer were comfort, he would step back.

14.3 — Quantum Mechanics Through Faraday's Eyes

Faraday would not reject quantum mechanics. He would reinterpret it quietly.

He would notice that quantum behavior appears most strongly when systems are highly constrained, isolated, or prevented from completing equilibrium in classical ways. He would see that measurement is not passive observation but reconfiguration—a boundary imposed on a system that forces redistribution of outcomes.

He would be intrigued, not confused, by non-local correlations. He never believed influence traveled as a messenger. He believed systems behaved as wholes when structure demanded it. Entanglement, to him, would not imply spooky action, but shared configuration awaiting completion.

What modern physics calls "collapse," Faraday might treat as forced resolution—a system compelled to finish in one way because boundaries no longer permit ambiguity. He would not ask where the wavefunction went. He would ask what changed in the system's ability to remain unresolved.

In this sense, Faraday would align naturally with interpretations that resist ontological excess and emphasize structure, constraint, and response. He would be deeply uncomfortable with any account that treats probability as ignorance rather than structural indeterminacy.

Most importantly, he would refuse to dramatize mystery.

Where modern discourse sometimes celebrates paradox, Faraday would treat paradox as a sign that language has outrun structure again. He would not try to resolve paradox with philosophy. He would return to experiment and ask: *What configuration makes this outcome unavoidable?*

14.4 — Engineering Success, Explanatory Drift

One of the great ironies Faraday would notice is this: engineering understands electromagnetism better than physics explains it.

Modern engineers design systems by respecting boundaries, managing impedance, shaping geometry, and controlling energy redistribution. They rarely speak of fields as substances. They speak of loads, constraints, pathways, and stability. Their language is closer to Faraday's instincts than to many textbook explanations.

Faraday would appreciate this deeply.

He would see that while ontology stagnated, practice evolved. That we learned how systems behave without fully agreeing on what they "are." That invention raced ahead of interpretation—and did so successfully.

But he would also sense the cost.

When explanation drifts too far from practice, education becomes ritualized. Students memorize rules without seeing necessity. They learn which formula applies where, but not why the system had no alternative. Physics becomes something to survive rather than understand.

Faraday believed understanding was moral, not decorative. To him, explanation mattered because it disciplined curiosity. He would worry that modern abstraction, when detached from structure, invites intellectual complacency—confidence without clarity.

14.5 — What Faraday Would Urge Us to Do Now

Faraday would not ask modern physics to abandon mathematics, technology, or abstraction. He would ask it to recover restraint.

He would urge us to separate three things that are currently entangled:

1. What we can calculate
2. What we can observe
3. What we choose to believe exists

He would insist that belief follow necessity, not convenience.

Faraday would push physics toward questions that sound simple but are transformative:

- *Where does equilibrium fail here?*
- *What boundary prevents completion?*
- *Why must the system respond in this way and not another?*

He would encourage fewer declarations and more re-descriptions. Less ontology, more structure. Fewer entities, more inevitability.

In short, he would ask physics to grow quieter.

Universal Energy Dynamics is aligned precisely with this demand. It does not diminish modern physics; it removes what modern physics no longer needs. It restores the explanatory direction Faraday trusted: from configuration to consequence, from denial to response, from structure to behavior.

If Faraday lived today, he would not call this revolutionary. He would call it overdue.

Closing Transition — Faraday as a Mirror, Not a Judge

Faraday would not judge modern physics harshly. He would recognize its achievements too well for that. But he would act as a mirror—reflecting back a question we have learned to postpone:

Do we understand this, or do we merely operate it?

This chapter does not imagine Faraday as a critic standing above modern science. It places him beside us, asking the same question he always asked—not *what can we say*, but *what are we justified in saying?*

The next chapter turns from Faraday's hypothetical gaze to our actual condition. It examines how invention flourished while understanding lagged, why ontology hardened even as practice advanced, and what it would mean to rebalance physics toward explanation without sacrifice.

From here on, the book no longer asks what Faraday would think. It asks what responsibility his discipline leaves us with now.

Chapter 15 — From Invention to Understanding

Why engineering succeeded while ontology stagnated

A Strange Asymmetry

Modern physics presents an unusual asymmetry.

We build machines of extraordinary reliability. We transmit power across continents, compute at quantum scales, image atoms, stabilize satellites, and manipulate matter with precision Faraday could never have imagined. And yet, when asked *what is actually happening*, explanations often retreat into metaphor, abstraction, or silence.

This chapter confronts that asymmetry directly.

It does not accuse physics of failure. On the contrary, it begins by acknowledging a difficult truth: physics has never been more successful—and never more uncertain about its own explanations. Engineering thrives. Prediction excels. Ontology hesitates.

The question is not *why technology advanced.* That part is obvious. The question is why understanding did not advance at the same pace.

15.1 — Why Engineering Could Advance Without Ontology

Engineering does not require metaphysical certainty. It requires repeatability under constraint.

An engineer does not need to know what a field "is" to design a motor. She needs to know how geometry, materials, boundaries, and loads shape behavior. She needs to know where instability appears, how it propagates, and how it can be controlled. This is structural knowledge, not ontological commitment.

In this sense, engineering matured by quietly following a Faraday-like path.

It learned to work with:

- impedance rather than force,
- boundary conditions rather than substances,
- stability criteria rather than causal stories,
- configuration rather than essence.

Engineering advanced because it trusted behavior over explanation. If a system behaved reliably when arranged correctly, that was enough. The question of *what truly exists* could be postponed indefinitely.

Physics, meanwhile, inherited a different responsibility. It was expected not only to predict, but to explain. And explanation, historically, was equated with naming entities.

This divergence explains the asymmetry. Engineering flourished because it did not need ontology. Physics struggled because it felt obliged to supply one.

15.2 — How Ontology Quietly Hardened

The entities of modern physics did not arrive through proof. They arrived through usefulness.

Fields, particles, wavefunctions, virtual excitations—each began as a mathematical or conceptual convenience. They allowed equations to close, predictions to stabilize, and reasoning to proceed.

Over time, repeated success lent them credibility. Credibility hardened into assumption.

What began as language slowly became belief.

This hardening was subtle. No one declared, "This metaphor is now reality." Instead, the language became habitual. Textbooks repeated it. Diagrams reinforced it. Generations learned to think in terms of entities rather than constraints.

Eventually, questioning ontology felt unnecessary—even obstructive.

The result was a peculiar situation: physics became extraordinarily precise while its explanations grew increasingly optional. When pressed, many physicists now respond that interpretation does not matter, only prediction does. This is not an error; it is a coping mechanism.

But it comes at a cost.

When ontology stagnates, understanding becomes fragile. Concepts cannot be transferred cleanly across domains. Students memorize instead of reason. Paradoxes multiply, not because nature is paradoxical, but because language is doing work it was never disciplined to do.

15.3 — The Educational Consequence

Nowhere is this imbalance more visible than in education.

Students are taught how to calculate long before they are taught how to *see*. They learn which equation applies to which problem, but not why the system had no alternative but to behave that way. Knowledge becomes procedural rather than structural.

This is why electromagnetism, quantum mechanics, and even classical mechanics often feel unintuitive—not because they are inherently complex, but because explanation has been outsourced to formalism.

The student is told:

- "This is how it is."
- "This is what the equation says."
- "Do not worry about meaning."

Faraday would have found this deeply troubling.

For him, understanding was not optional decoration; it was the discipline that prevented self-deception. Without understanding, success becomes brittle. You can operate a system without knowing when it will fail—but not why.

The stagnation of ontology thus shows up not as theoretical crisis, but as pedagogical exhaustion.

15.4 — Why Physics Defended the Status Quo

It is important to understand that this stagnation was not negligence.

Physics defended abstraction because abstraction worked.

When equations predict perfectly, questioning interpretation feels unnecessary. Worse, it feels dangerous. Why risk destabilizing a successful framework? Why reopen foundational questions when technology depends on settled assumptions?

This is why reinterpretation is often resisted—not on empirical grounds, but on pragmatic ones. The fear is not that a new explanation is wrong, but that it is unnecessary.

Universal Energy Dynamics challenges this fear quietly.

It does not alter equations. It does not disturb predictions. It does not demand new experiments to justify itself. It asks only that explanation be realigned—from entities acting, to structures constraining behavior.

This is not a threat to physics. It is a relief.

15.5 — The Missed Opportunity After Faraday

After Faraday, physics stood at a crossroads.

One path led toward formalism without restraint—successful, powerful, but conceptually thin. The other path required a language of structure that did not yet exist. Physics chose the first, because the second was not available.

That choice made modern science possible.

But it also delayed a necessary return.

The opportunity missed was not to invent better entities, but to ask a better question: *What does the system forbid?* That question does not require new mathematics. It requires intellectual humility.

Universal Energy Dynamics represents that return—not backward, but inward.

15.6 — Understanding as the Next Evolution

Understanding is not a luxury phase that follows invention. It is the stabilization phase that prevents invention from becoming ritual.

Physics has reached a point where prediction is no longer the limiting factor. Explanation is. We do not need more metaphors. We need fewer, better disciplined ones—or none at all.

Understanding, in this sense, means:

- explaining without inventing,
- naming only what necessity demands,
- trusting structure over story.

This is not a rejection of abstraction. It is a reorientation of it.

Just as engineering matured by learning to respect constraints rather than substances, physics can mature by learning to explain inevitability rather than assert existence.

Closing Transition — From Power to Clarity

Engineering taught us how to make things work. Physics must now remember how to make them make sense.

This chapter does not argue that ontology must be abolished. It argues that ontology must be earned—slowly, reluctantly, and only when no structural explanation remains.

Faraday understood this instinctively. He chose silence over invention. Modern physics chose invention over silence. Both choices were historically necessary.

Now, a third choice becomes possible.

We can keep the power of our equations, the reach of our technology, and the success of our predictions—while restoring explanation to its proper discipline. Not by adding more entities, but by removing what explanation no longer needs.

The final chapter closes this book by returning to the simplest truth Faraday left us with:

He did not discover fields. He discovered change.

And restoring that insight—not as nostalgia, but as method—may be the next evolution of physics itself.

Closing Chapter

Why restoring intuition is the next evolution of physics

Faraday Didn't Discover Fields, He Discovered Change

History remembers Michael Faraday as the father of fields. That attribution is understandable—and wrong.

Faraday never claimed to have discovered a new substance filling space. He never insisted that "lines of force" were entities acting independently. He did not seek to populate the universe with invisible actors. What he did—consistently, relentlessly, and almost stubbornly—was something far more radical:

He treated change itself as the primary fact of physics.

Fields were never his destination. They were scaffolding. Visualization aids. Temporary language used to discipline intuition where mathematics was unavailable. The tragedy is not that Faraday lacked formalism—it is that later generations mistook his visual discipline for ontology.

This final chapter closes the book by restoring the original center of gravity: not fields, not forces, not particles—but change constrained by structure.

What Faraday Actually Observed

Faraday's experiments never asked, *What exists?* They asked, *What must happen when this configuration is altered?*

Whether it was induction, rotation, shielding, or redistribution, the pattern was consistent:

- A system at rest behaves predictably
- A system disturbed seeks resolution
- When resolution is blocked, motion persists
- When boundaries change, behavior reorganizes

Nothing in this requires a field as an entity. Everything requires asymmetry, constraint, and incomplete equilibrium.

Faraday noticed something subtle but decisive: nature responds to change, not to labels.

A magnet sitting quietly does nothing remarkable. A current held steady produces no novelty. It is only when something changes—position, orientation, connection, geometry—that physics reveals itself.

This is not poetic framing. It is empirical.

How Fields Became a Conceptual Shortcut

The idea of fields emerged not because Faraday demanded it, but because physics needed a way to talk about continuity without invoking action-at-a-distance.

Fields solved a linguistic problem. They were never meant to solve an ontological one.

Over time, however, the shortcut hardened. Lines of force became field lines. Field lines became fields. Fields became entities. And once entities exist, they begin to act, store energy, propagate influence, and demand interpretation.

At that moment, physics quietly inverted Faraday's insight.

Instead of saying *change produces effects*, we began saying *fields act on charges*. Instead of configuration driving motion, we described agents pushing outcomes. The language shifted from necessity to causation.

This shift was not malicious. It was efficient. But it had consequences.

Once fields became actors, intuition lost its anchor. Explanations multiplied. Metaphors stretched. Students learned what to calculate, not what to see.

Faraday would not have objected loudly. He would have gone silent.

Why Change Is More Fundamental Than Force

Force presumes agency. Change requires none.

Force implies something doing something to something else. Change simply describes that a system cannot remain as it is given its constraints. This distinction matters profoundly.

Under Universal Energy Dynamics, Faraday's experiments resolve cleanly:

- Induction occurs because energy distribution cannot remain balanced when configuration changes
- Motion persists because equilibrium is denied, not because force is applied
- Shielding works because boundaries redirect redistribution, not because fields are excluded
- Stability arises when all available resolution paths are exhausted

In every case, change is not an effect—it is the trigger.

This is why Faraday's work scales so effortlessly from motors to transformers to eddy currents to modern systems. He discovered a logic that does not depend on scale, substance, or interpretation.

Change is universal. Resolution is inevitable. Equilibrium is conditional.

Everything else is vocabulary.

The Cost of Losing Intuition

Physics did not lose intuition overnight. It traded it—slowly—for efficiency.

Mathematics offered certainty. Prediction offered authority. Engineering offered validation. Over time, intuition was treated as optional, even dangerous. Why rely on visualization when equations never fail?

But intuition is not decoration. It is error control.

Without intuition:

- We cannot tell when language overreaches
- We cannot distinguish metaphor from mechanism
- We cannot explain without inventing
- We cannot teach without ritual

Faraday understood this instinctively. He refused explanations he could not visualize—not because they were wrong, but because they were undisciplined.

Restoring intuition does not mean abandoning rigor. It means restoring alignment between language and necessity.

UED as Continuation, Not Replacement

Universal Energy Dynamics does not correct Faraday. It completes what language could not yet carry.

Faraday saw configuration-driven behavior but lacked a neutral vocabulary to describe it without metaphor. Fields filled that gap temporarily. UED removes the scaffolding now that the structure can stand.

This framework:

- Retains all empirical success
- Discards unnecessary ontology
- Re-centers explanation on structure and delta
- Treats energy as redistribution, not substance
- Treats time as resolution order, not dimension

Nothing is lost. Much is clarified.

UED does not ask physics to restart. It asks physics to clean up after itself.

What the Next Evolution of Physics Requires

The next evolution of physics will not come from:

- more particles
- deeper abstractions
- larger machines
- stranger metaphors

It will come from conceptual restraint.

From asking, again and again:

- *What is actually changing here?*
- *What prevents resolution?*
- *Why can the system not settle?*

These are Faraday's questions. They are still sufficient.

The future of physics depends less on discovery and more on interpretive honesty. On the courage to say, "This explains behavior without inventing entities—and that is enough."

Final Resolution — Faraday's Quiet Victory

Faraday did not lose to formalism. He outlived it.

Every time physics struggles with interpretation—whether in electromagnetism, quantum mechanics, or cosmology—it circles back to the same realization: behavior precedes explanation.

Faraday knew this before equations made it fashionable to forget.

He did not discover fields. He did not invent forces. He did not populate space with invisible actors.

He discovered something far more enduring:

When change is constrained, nature must respond—and that response is physics.

Restoring that insight is not nostalgia. It is maturity.

And with that, this book does not close an argument. It restores a discipline.

Appendices

The past does not instruct us by being correct. It instructs us by showing where restraint once protected truth.

Appendix A — Clean UED Definitions

Operational clarity without ontological inflation

For Electromagnetism

This appendix fixes the working meanings of key terms as they are used throughout this book. These are *operational definitions*, not metaphysical claims. They exist to discipline explanation, prevent metaphor drift, and keep language aligned with structure and necessity.

None of the definitions below assert what *ultimately exists*. They specify what must be assumed, constrained, or denied for electromagnetic behavior to become unavoidable.

A.1 Energy Configuration

What this is not: Energy configuration is not a substance distributed in space, not a stored entity, and not a property owned by objects.

UED definition: Energy configuration refers to the **arrangement of a system that determines how energy redistribution is permitted or restricted**. It includes geometry, connectivity, orientation, material constraints, and boundary conditions that together define which equilibrium paths are available and which are denied.

Under UED, energy does not "sit" in a configuration. Rather, the configuration qualifies how redistribution must occur when change is introduced.

Why this framing matters: This removes the need to imagine energy as something residing in fields or components. Behavior follows from arrangement, not possession.

A.2 Delta (Δ)

What this is not: Delta is not a force, not a fluctuation, not noise, and not merely a numerical difference in measured values.

UED definition: Delta is the **structural non-equivalence introduced into a system**—a mismatch between what the system is configured to allow and what completion (equilibrium) would require. Delta arises from change: motion, switching, rotation, boundary alteration, or reconfiguration.

Delta is the condition that makes response unavoidable.

Why this framing matters: By treating delta as structural rather than energetic or probabilistic, UED explains why change—not magnitude—is the decisive factor in induction and motion.

A.3 Coupling

What this is not: Coupling is not interaction via a mediating substance, not a transmission channel, and not an invisible link acting at a distance.

UED definition: Coupling refers to the **degree to which two parts of a system share redistribution constraints**. Two regions are coupled when a change in one necessarily alters the completion conditions of the other.

Coupling is therefore a structural dependency, not an action.

Why this framing matters: This allows electromagnetic influence without invoking fields "acting" across space. Effects propagate because constraints propagate.

A.4 Equilibrium

What this is not: Equilibrium is not rest, not inactivity, and not the absence of energy.

UED definition: Equilibrium is the **state in which all permitted redistribution paths are satisfied**. A system is in equilibrium when no unresolved delta remains given its current configuration.

Importantly, equilibrium is conditional. It depends entirely on configuration and boundaries.

Why this framing matters: This explains why some systems settle quickly while others sustain motion or current indefinitely: equilibrium is denied by structure, not resisted by force.

A.5 Boundary

What this is not: A boundary is not merely a surface, shield, or container, and not a passive separator.

UED definition: A boundary is any **structural condition that restricts, redirects, or completes redistribution paths**. Boundaries may be physical (conductors, insulators), geometric (loops, gaps), or configurational (constraints on motion or connection).

Boundaries do not block energy; they shape inevitability.

Why this framing matters: This clarifies phenomena such as shielding, confinement, and isolation without appealing to field exclusion or force cancellation.

A.6 Relationship Between the Terms (Summary)

Under Universal Energy Dynamics:

- **Energy configuration** defines what redistribution is possible
- **Delta** arises when configuration and completion conflict

- **Coupling** determines how that conflict propagates
- **Equilibrium** is the resolution of all permitted paths
- **Boundaries** decide which resolutions are allowed or forbidden

Electromagnetic phenomena emerge *not because entities act*, but because systems *cannot finish equilibrium* under imposed arrangements.

A.7 Scope Reminder

These definitions are:

- Context-specific to electromagnetism
- Compatible with existing formalism
- Neutral with respect to metaphysical claims

They exist to *anchor explanation*, not to close inquiry.

Used consistently, they allow the reader to translate every phenomenon discussed in this book—from induction to motors to shielding—back into structure, necessity, and response, without invoking unnecessary actors.

This appendix therefore functions as a conceptual reference plane. Nothing in the main text should require terms beyond what is defined here.

Appendix B — Classical Laws Rewritten in UED Language

Preserving prediction while removing force metaphors

This appendix does not restate classical laws to improve their accuracy. Their predictive success is already proven. The purpose here is narrower and more disciplined:

To show that the same laws remain fully operational when rewritten without force-centric or entity-based explanations, and that their behavior follows directly from configuration, delta creation, and equilibrium resolution.

No equations are replaced. No experimental claims are altered. Only the direction of explanation is corrected.

B.1 Faraday's Law — Induction Without "Flux Acting"

Canonical statement (standard form)

A changing magnetic flux through a closed loop induces an electromotive force (EMF) in the loop.

What it predicts operationally

- A current appears when the system is reconfigured by motion, switching, or time-variation
- No current appears when the configuration is static
- The magnitude depends on the rate of change

What UED removes

- The idea that "flux cuts conductors"
- The notion that a magnetic field *acts on* charges
- Any requirement for a field entity doing causal work

UED reinterpretation

Under UED, induction occurs because a closed conductive configuration cannot maintain equilibrium when its redistribution constraints are altered.

Change (motion, switching, rotation) introduces delta—a structural mismatch between allowed completion paths and current configuration. The induced EMF is the system's only available resolution.

No agent induces current. The system resolves inevitability.

B.2 Lenz's Law — Direction Without Opposition

Canonical statement (standard form)

The induced current flows in a direction that opposes the change producing it.

What it predicts operationally

- Induced currents resist rapid change
- Systems exhibit stabilizing behavior
- Energy is conserved without runaway acceleration

What UED removes

- The anthropomorphic idea of "opposition"
- Any suggestion of intent or resistance
- The force-reaction framing

UED reinterpretation

Under UED, induced current flows in the direction that reduces unresolved delta most efficiently given existing constraints.

The system does not oppose change. It selects the only resolution path that minimizes structural imbalance.

Lenz's Law is therefore not reactive—it is equilibrium-seeking inevitability.

B.3 Lorentz Force Law — Motion Without Push

Canonical statement (standard form)

A charge moving in electric and magnetic fields experiences a force proportional to its velocity and field strength.

What it predicts operationally

- Deflection of charge trajectories
- Curved paths in magnetic environments
- Velocity-dependent behavior

What UED removes

- The concept of a force acting on a charge
- The assumption of fields as active agents

- Push-pull causation

UED reinterpretation

Under UED, charged motion occurs because the system's configuration does not permit straight-line equilibrium completion.

Velocity introduces directional asymmetry within constrained geometry. The resulting trajectory is not forced—it is the only permitted completion path under the imposed arrangement.

Motion curves because straight resolution is denied.

B.4 Faraday Disk (Unipolar Induction) — Why Flux Language Fails

Canonical problem

Current is generated even when magnetic flux through the circuit does not change.

What it predicts operationally

- EMF appears during rotation
- Classical flux-cutting explanations fail
- Textbooks resort to patch explanations

What UED removes

- Flux as causal mechanism
- Field-line cutting narratives
- Frame-dependent confusion

UED reinterpretation

Rotation alters redistribution geometry, not flux quantity.

The conducting disk continuously creates delta along its radius due to differential motion under constraint. EMF arises because equilibrium cannot be globally satisfied across the rotating structure.

The paradox disappears once change—not flux—is treated as primary.

B.5 Faraday Cage — Shielding Without Exclusion

Canonical statement (standard form)

A conductive enclosure shields its interior from external electric fields.

What it predicts operationally

- Interior remains stable
- External disturbances do not penetrate
- Charges redistribute on the surface

What UED removes

- The idea that fields are blocked or cancelled
- Field exclusion metaphors
- Shielding as negation

UED reinterpretation

The enclosure provides boundary completion.

Redistribution resolves entirely on the exterior because the boundary permits full equilibrium resolution there. The interior remains stable not because influence is blocked, but because no delta remains to propagate inward.

Stability is achieved by completion, not protection.

B.6 Why No Law Is Broken Under UED

Across all cases:

- Predictions remain identical
- Mathematical formalisms remain usable
- Engineering outcomes remain unchanged

What changes is only explanation.

Classical laws continue to function because they already encode correct behavior. UED simply removes unnecessary actors and restores causal direction to structure.

B.7 What This Appendix Demonstrates

This appendix demonstrates three critical points:

1. Force metaphors are optional
2. Entities are not required for prediction
3. Structural inevitability explains all classical electromagnetic behavior

Nothing needs to be overturned. Nothing needs to be replaced.

Explanation simply becomes quieter, cleaner, and more faithful to Faraday's discipline.

This completes the translation layer between classical electromagnetism and Universal Energy Dynamics—without rewriting physics, and without inventing what behavior does not demand.

Appendix C — What This Book Is Not Claiming

Scope boundaries and intellectual honesty

This appendix exists to prevent misreading. The arguments in this book are precise by design, but precision is often mistaken for provocation. To avoid that, the boundaries below are stated explicitly.

What follows are negations with purpose. Each clarifies what this work intentionally avoids, and why that restraint matters for intellectual integrity.

C.1 This Book Is Not Claiming That Classical Electromagnetism Is Wrong

This work does not dispute the correctness of classical electromagnetic laws, their mathematical form, or their predictive success.

Why this matters: The project is interpretive, not corrective. Prediction already works; explanation needed discipline.

C.2 This Book Is Not Proposing New Equations or Modifying Existing Ones

No new laws, constants, formalisms, or computational frameworks are introduced.

Why this matters: UED operates at the level of explanation and ontology restraint, not mathematical replacement.

C.3 This Book Is Not Rejecting Maxwell's Equations

Maxwell's equations remain valid, powerful, and indispensable.

Why this matters: The critique concerns how those equations are interpreted, not their operational role.

C.4 This Book Is Not Claiming That Fields Do Not Exist

This work does not assert that fields are unreal, illusory, or false.

Why this matters: UED refuses to make metaphysical claims where structural explanation already suffices.

C.5 This Book Is Not Arguing That Engineering Is Conceptually Inferior

Engineering success is treated as evidence of deep behavioral understanding.

Why this matters: The problem is not practice—it is the gap between practice and explanation.

C.6 This Book Is Not Reinterpreting Quantum Mechanics in Full

Quantum phenomena are referenced only where interpretive parallels are unavoidable.

Why this matters: A full quantum reinterpretation requires separate treatment and is beyond this book's scope.

C.7 This Book Is Not Offering a Final Ontology of Reality

No claim is made about the ultimate nature of space, energy, matter, or existence.

Why this matters: UED is a discipline of explanation, not a metaphysical closure.

C.8 This Book Is Not Anti-Mathematics or Anti-Abstraction

Formalism is respected as stabilizing, predictive, and essential.

Why this matters: The concern is not abstraction itself, but abstraction detached from necessity.

C.9 This Book Is Not Claiming Michael Faraday "Secretly Knew" UED

Faraday is not retrofitted with modern terminology or hidden intent.

Why this matters: The alignment is methodological, not historical or psychological.

C.10 This Book Is Not a Polemic Against Modern Physics

No school, individual, or tradition is attacked or dismissed.

Why this matters: The aim is clarification, not confrontation.

C.11 This Book Is Not Declaring the End of Physics

No claim is made that physics has reached completion or closure.

Why this matters: UED restores explanatory hygiene, not final answers.

C.12 What This Book Is Claiming

This book claims only this:

Electromagnetic phenomena can be fully explained through energy configuration, delta creation, coupling, equilibrium resolution, and boundary conditions, without requiring force-centric or entity-based metaphors—and that this explanation aligns more faithfully with Faraday's discipline.

Nothing more. Nothing less.

C.13 Final Boundary Statement

If the reader disagrees with this book, the disagreement should address explanatory necessity, not historical loyalty or mathematical success.

The argument lives or fails on one criterion alone:

Does removing unnecessary ontology reduce clarity—or restore it?

This appendix exists so that debate, if it arises, remains about what the book actually says—and not about claims it deliberately refuses to make.

For Physicists and Engineers

This book is not asking you to relearn electromagnetism. It is asking you to reorient how you explain what you already know works.

If you approach it with that distinction in mind, the reading becomes precise, efficient, and—ideally—useful.

If You Are a Physicist

Read this book concept-first, not equation-first.

Nothing here challenges the validity of Maxwellian formalism, boundary-value methods, or predictive frameworks. You will not be asked to abandon fields, operators, or established calculations. What *is* challenged is the automatic slide from mathematical success to ontological certainty.

As you read:

- Treat equations as *given and trusted*
- Treat explanations as *open to discipline*
- Notice where language introduces agents, entities, or actions that experiments never demand

When the text reframes familiar laws, resist the urge to ask, *"Is this equivalent mathematically?"* Instead, ask the Faraday question:

Does this explanation rely on anything that the behavior itself does not force me to assume?

This book rewards readers who are comfortable separating:

- prediction from explanation
- formalism from interpretation
- usefulness from necessity

If you find yourself thinking, *"This removes something I never really needed anyway,"* you are reading it correctly.

If You Are an Engineer

Read this book recognition-first, not defense-first.

Much of what is described here will feel familiar—not because you learned it this way in textbooks, but because you already work this way in practice. You design by respecting constraints. You trust geometry, boundaries, impedance, stability, and failure modes more than metaphysical stories.

As you read:

- Map ideas directly onto systems you already understand
- Notice how little depends on invisible agents
- Observe how explanation aligns with how machines actually behave

You are not being asked to become philosophical. You are being given a language that legitimizes the intuition you already rely on, but rarely see acknowledged in theoretical discourse.

If you find yourself thinking, *"This explains why my designs work without invoking extra stories,"* you are reading it correctly.

If You Are Both (or Moving Between Them)

This book is written precisely for that overlap.

It occupies the space where:

- equations are respected

- intuition is restored
- explanation stops inventing

Read it slowly, but not defensively. The arguments are cumulative, not confrontational. Early chapters establish discipline; later chapters rely on it. Skipping ahead is possible, but coherence improves when the structure is allowed to build.

What This Book Will Not Ask of You

- It will not ask you to reject successful physics
- It will not ask you to adopt a new mathematical framework
- It will not ask you to defend Universal Energy Dynamics as ideology

It asks only this:

When two explanations predict the same behavior, prefer the one that assumes less, not more.

That principle is not new. It is the discipline practiced instinctively by Faraday—and quietly relied upon by generations of engineers.

This book simply restores it to explicit use.

A Final Reading Hint

If at any point you feel the explanation becoming *simpler*, not *weaker*, pause there.

That simplicity is not reduction. It is restraint.

And restraint, in physics, is often the most advanced move available.

Closing Note — In Place of Certainty

I did not write this book to challenge physics. I wrote it to challenge how comfortably we explain what already works.

Faraday has been with me throughout this journey—not as a historical figure to be admired, but as a discipline to be respected. He never rushed to name what he could not justify. He never allowed language to outrun observation. And when explanation threatened to become invention, he chose silence over certainty.

That choice stayed with me.

Modern physics is powerful beyond anything Faraday could have imagined. Our equations are precise. Our technologies are astonishing. Our predictions are reliable. And yet, something subtle has drifted. We explain too quickly. We trust metaphors too easily. We accept entities where structure would suffice.

Universal Energy Dynamics did not emerge from a desire to replace what exists. It emerged from a discomfort that many of us carry quietly—the feeling that we *understand how things behave*, but are less certain that we understand why we say what we say about them.

This book is an attempt to restore alignment.

Not alignment between theories, but alignment between:

- behavior and explanation
- success and honesty
- intuition and restraint

I have deliberately avoided grand claims. I have resisted final ontologies. I have chosen completion without correction, clarity without confrontation. If at times this feels unsatisfying, that discomfort is intentional. Understanding should sharpen curiosity, not anesthetize it.

Faraday did not give us answers wrapped in certainty. He gave us a method that refused to lie.

If this book has done anything worthwhile, it is this: it has shown that we can explain deeply without adding more than nature demands, and that sometimes the most advanced move in science is not invention—but restraint.

Where this work goes next is not fixed. That is as it should be. What matters is that explanation remains disciplined, language remains honest, and curiosity remains accountable to structure.

If you finish this book seeing electromagnetism a little more quietly—but a little more clearly—then it has done its job.

— **Sandeep Chavan**

Acknowledgments

This book owes its existence less to agreement than to **resistance**—the quiet, productive resistance that arises when something familiar no longer feels fully honest.

I am grateful to **Michael Faraday**, not in the ceremonial sense reserved for historical giants, but in the practical sense reserved for teachers who continue to correct us long after their time. His discipline—especially his refusal to explain what he could not justify—has been the silent standard against which every page of this book was measured.

I owe thanks to the generations of **engineers** who, often without philosophical language, preserved Faraday's intuition in practice. Their designs, constraints, and failure analyses understand electromagnetism more honestly than many explanations ever did. This book recognizes that debt openly.

I am equally indebted to **physicists** who, even while working within powerful formalisms, remain uneasy with easy stories. Those who sense the gap between calculation and explanation—but keep asking anyway—are the quiet custodians of scientific integrity. This work belongs to that lineage.

Some acknowledgments are indirect but essential. To the students who asked simple questions that textbooks answered too quickly. To the skeptics who forced precision by refusing comfort. To the traditions of science that valued restraint as much as creativity—even when that value went temporarily out of fashion.

Finally, I acknowledge the discipline of *stopping*.

Stopping before naming. Stopping before asserting. Stopping before turning explanation into belief.

This book exists because someone once stopped at the edge of certainty—and because we are now finally ready to understand why that pause mattered.

— **Sandeep Chavan**

About the Author

Sandeep Chavan is an independent researcher, educator, and author whose work bridges physics, philosophy, and systems-level thinking. Trained in engineering and deeply engaged with foundational questions in science, his writing focuses on restoring conceptual clarity where formal success has outpaced explanation.

He is the originator of **Universal Energy Dynamics (UED)**—a framework that emphasizes configuration, constraint, equilibrium, and necessity over force-centric or entity-based narratives. His work does not seek to replace established formalisms, but to discipline interpretation by aligning language with observable structure and behavior.

As a teacher, Sandeep is known for prioritizing intuition, restraint, and first-principle reasoning, especially in areas where students and practitioners feel a disconnect between calculation and understanding. As an author, he writes to clarify—not to provoke—favoring precision over polemic and completion without correction.

This book reflects his broader intellectual commitment: to preserve the power of modern science while recovering the honesty that allows explanation to remain faithful to what nature actually demands—and no more.

Further Reading

From UED and Chavanian Frameworks

For readers who wish to go deeper without expanding scope prematurely

The works listed below extend the ideas used in this book without repeating it. They are organized to preserve the same discipline practiced here: explanation before ontology, structure before metaphor, and restraint before closure.

None of these texts are prerequisites. They are continuations—each exploring one dimension that this book deliberately touched only as much as necessary.

Foundational UED Frameworks

1. Universal Energy Dynamics (UED): A Structural Interpretation of Physical Behavior: Explores the core logic of UED in a domain-neutral form. Introduces energy as redistribution, equilibrium as conditional, and configuration as the primary explanatory driver—without focusing on electromagnetism specifically.

2. The Still Universe: Why Motion Is a Resolution, Not a Property: Develops the idea that motion arises only when equilibrium is denied. This work provides deeper grounding for the non-equilibrium reasoning used throughout the Faraday reinterpretation.

3. Delta, Alignment, and Resolution: A focused treatment of delta as a structural condition rather than a force or fluctuation.

Particularly useful for readers interested in induction, stability, and irreversibility across systems.

Electromagnetism and Applied Domains

4. Electromagnetism Without Fields: A technical companion that extends the arguments of this book into engineering-relevant contexts—motors, transformers, shielding, and signal systems—using UED language exclusively.

5. Boundary Physics: Why Surfaces Decide Behavior: Examines boundaries as active resolution agents rather than passive separators. This work deepens the logic behind Faraday cages, confinement, and stability.

Time, Causality, and Non-Equilibrium

6. The Illusion of Time: Causality as Resolution Order: Explores time as a consequence of unresolved redistribution rather than a background dimension. Readers interested in how UED reframes temporal sequence will find this essential.

7. Entropy Without Disorder: Reinterprets entropy as structural exhaustion rather than randomness. This book provides conceptual continuity between thermodynamics and UED's equilibrium logic.

Mind, Observation, and Interpretation (Chavanian Frameworks)

8. Awareness, Alignment, and the Observer: Examines observation as a configurational act rather than passive measurement. Particularly relevant for readers thinking about quantum interpretation without metaphysical escalation.

9. The Discipline of Not Explaining Too Much: A philosophical work addressing why premature explanation damages understanding. This book complements Faraday's methodological restraint with broader epistemic insights.

For Readers Crossing Disciplines

10. Engineering Intuition as First-Principle Physics: Bridges practical engineering reasoning with foundational physics, showing how constraint-based thinking often outperforms entity-based explanations in real systems.

Reading Guidance

These works are *modular*, not sequential. Readers are encouraged to follow curiosity rather than order. If one principle guided this book, it applies here as well:

Read only as far as explanation remains necessary. Stop when structure feels sufficient.

That restraint is not a limitation—it is the method.

Aphorisms & Reflections

Quiet statements distilled from the discipline of the book

These aphorisms are not conclusions. They are **stopping points**—places where explanation becomes clean enough to pause.

They are meant to be reread slowly, not agreed with quickly.

On Explanation

- Understanding improves not when we add answers, but when we remove unnecessary ones.
- Explanation becomes dishonest the moment it outruns necessity.
- Clarity is not achieved by saying more, but by stopping earlier.
- If a system behaves inevitably, explanation should sound inevitable too.
- The best explanations feel smaller after you accept them.

On Change and Necessity

- Change is not dramatic. It is decisive.
- Nature does not respond to causes; it responds to unresolved conditions.
- When equilibrium is denied, motion is not optional.
- Persistence is not energy—it is unfinished resolution.
- What continues is not what wants to, but what cannot yet

finish.

On Fields, Forces, and Language

- Fields solved a language problem before they became a belief problem.
- Force is often what we name when we forget to ask what configuration forbids.
- Metaphors are useful until they begin to explain themselves.
- When language acts, understanding retreats.
- If removing an entity changes nothing observable, it was never required.

On Boundaries and Structure

- Boundaries do not stop behavior; they decide where it completes.
- Stability is achieved by completion, not by protection.
- What appears isolated is often simply resolved.
- Geometry is not decoration—it is instruction.
- Systems behave according to what they are not allowed to do.

On Mathematics and Meaning

- Equations can be correct while explanations drift.
- Mathematics stabilizes prediction; it does not guarantee truth.
- Symbols are safe tools until they are mistaken for reality.
- Precision without restraint becomes ritual.
- When meaning disappears, calculation quietly takes over.

On Engineering and Practice

- Machines work because structure does not lie.
- Engineering understands necessity even when language hesitates.
- Practice often knows what theory has not yet learned how to say.
- Design succeeds by respecting limits, not by denying them.
- What fails teaches structure faster than what works.

On Faraday's Discipline

- Faraday stopped where honesty required silence.
- He did not explain less because he knew less, but because he refused to invent.
- His greatest discovery was not a phenomenon, but a method.
- He trusted repetition more than reputation.
- He changed physics by changing how attention was applied. *(Michael Faraday)*

On Universal Energy Dynamics (UED)

- UED explains by subtraction.
- Nothing essential is added; much unnecessary is removed.
- Energy does not travel—it redistributes when it must.
- Equilibrium is not rest; it is completion.
- Explanation becomes powerful when it becomes quiet.

On the Future of Understanding

- The next advance in physics will not be louder.
- It will use fewer words, not bigger ones.

- Progress is not adding layers, but learning when to stop.
- We no longer lack tools—we lack restraint.
- Maturity in science looks like calm.

Final Reflection

- When explanation feels inevitable, belief becomes unnecessary.

Disclaimer and Statement of Originality

Disclaimer

This book is a work of interpretation and conceptual clarification. It does **not** propose new physical laws, modify existing equations, or dispute the empirical success of established theories in electromagnetism or physics more broadly.

All scientific references—historical, experimental, and theoretical—are discussed in good faith and for explanatory purposes only. Any reinterpretation presented here concerns **language, ontology, and explanatory framing**, not predictive validity or experimental reliability.

The views expressed are the author's own and should not be construed as representing institutional positions, academic consensus, or formal endorsements. Readers are encouraged to engage critically, independently, and responsibly with the material.

This book is not intended as a substitute for formal scientific education, professional engineering guidance, or technical training. Its purpose is conceptual—to improve clarity in how physical behavior is explained, not to prescribe methods for design, experimentation, or application.

Statement of Originality

The framework referred to in this book as **Universal Energy Dynamics (UED)** is the original intellectual work of the author.

While it draws upon well-established physical principles, historical experiments, and widely known scientific results, the **interpretive structure, terminology, and explanatory discipline** presented here are original.

This book does not retroactively assign modern theories or intentions to historical figures. In particular, any alignment drawn between Michael Faraday's methodology and UED is **methodological**, not historical or psychological. Faraday's work is treated with respect and without speculative attribution.

The manuscript has been written independently and reflects the author's own synthesis, reasoning, and conclusions. Any similarities to existing interpretations are coincidental or arise naturally from engagement with shared scientific material. Where concepts overlap with established ideas, they are reframed to serve clarity rather than novelty.

No part of this book has been generated to imitate or replicate another author's unique expression. Quotations, when present, are limited, clearly contextualized, and used in accordance with fair use principles.

Closing Note on Responsibility

Originality in this work lies not in invention, but in **restraint**.

The author claims responsibility for the framework presented, but not authority over its acceptance. Readers are invited to test its clarity against experience, its discipline against habit, and its necessity against alternative explanations.

If this book encourages more careful language, fewer assumptions, and deeper attention to structure, it has fulfilled its intent.

Don't miss out!

Visit the website below and you can sign up to receive emails whenever Sandeep Chavan publishes a new book. There's no charge and no obligation.

https://books2read.com/r/B-A-EPGPC-CPKYI

Did you love *Faraday Reimagined*? Then you should read *What Einstein Did Exactly*[1] by Sandeep Chavan!

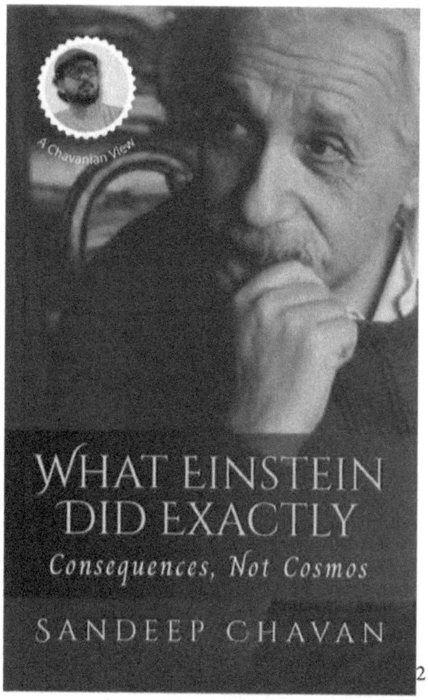

[2]

What if the greatest achievements of modern physics were never meant to explain reality—but only to describe its behavior?

What Einstein Did Exactly by Sandeep Chavan is a philosophical exploration of the boundaries between explanation and understanding, between mathematical description and physical meaning. Rather than proposing a new theory or challenging established science, this book asks a quieter but deeper question: what have we actually explained when we say we understand the universe?

Drawing on ideas from physics, philosophy, and the history of scientific thought, this work examines how modern science came

1. https://books2read.com/u/31w7QW

2. https://books2read.com/u/31w7QW

to rely on descriptive frameworks—models that predict with extraordinary accuracy, yet remain silent about underlying causes. It traces how concepts such as space, time, curvature, and expansion became accepted as physical realities rather than interpretive tools, and how this shift shaped modern scientific thinking.

In this book Chavan does not argue against Einstein or modern physics. Instead, he places Einstein's work in its proper intellectual context, showing how his restraint and precision allowed science to progress without claiming ultimate explanations. From this foundation, the text explores how later interpretations extended description into assumption, often without acknowledging the transition.

Rather than offering an alternative theory, this book invites readers to reconsider the boundary between knowledge and interpretation. It explores why certain questions persist despite technological progress, why explanations can be correct yet incomplete, and how meaning emerges not from answers alone, but from the structure of inquiry itself.

Written in clear, accessible language, *What Einstein Did Exactly* is intended for readers interested in physics, philosophy, and the nature of understanding—without requiring technical expertise. It is not a rejection of science, but a reflection on how science thinks, and what lies just beyond its current horizon.

Read more at https://www.sandeepjchavan.com/.